T0222516

Notable Modern Indian Mathematicians and Statisticians

Purabi Mukherji

Notable Modern Indian Mathematicians and Statisticians

During the 19th and 20th Centuries of Bengal

 Springer

Purabi Mukherji
Kolkata, West Bengal, India

ISBN 978-981-19-6131-1   ISBN 978-981-19-6132-8   (eBook)
https://doi.org/10.1007/978-981-19-6132-8

This Springer imprint is published by the registered company Springer Nature Singapore Pte Ltd.
The registered company address is: 152 Beach Road, #21-01/04 Gateway East, Singapore 189721,
Singapore

I dedicate this book to my father
Late Sushil Kumar Banerji,
a noted applied mathematician and
statistician, who inculcated in me the love for
mathematics and history of science.

Foreword

I am extremely happy to go through this wonderful monograph titled *Notable Modern Indian Mathematicians and Statisticians: During the 19th and 20th Centuries of Bengal*, by Dr. Purabi Mukherji. This book is a very informative piece of research done by an experienced researcher in the history of modern science in India. Dr. Purabi Mukherji has already published two books in the same field. The books were published by Springer Nature.

There have been remarkable and significant contributions in Mathematics and Statistics from this part of the world about which many of us may not be aware. Also, the current generation should know the work that was carried out by the pioneers from Bengal, under difficult circumstances, and be proud about it. This book is going to enlighten everyone. The personalities talked about here are pioneers in their own fields, and Dr. Mukherji has captured the exhaustive and chronological history very well. Their contributions are lucidly described, so that it is easy to read and understand by everyone. The compiled bibliographies will be very useful for future researchers in this area.

I convey my heartfelt congratulations to Dr. Mukherji for her sincere and well-researched efforts.

Prof. Bimal Kumar Roy
Indian Statistical Institute
Kolkata, West Bengal, India

Chairman, National Statistical
Commission
Government of India
New Delhi, India

Preface

This book portrays the lives and works of 16 pioneer mathematical scientists, who adorned 2 famous institutions of Calcutta—namely the University of Calcutta and the Indian Statistical Institute (ISI) and the first Indian Institute of Technology (IIT), elaborated in lucid language. They worked under very difficult circumstances. They did not have access to any modern gadgets, not even calculators. Paucity of funds made foreign journals and books almost out of bounds for them. But in spite of such impediments, their research works made a mark in the world of science, and they are critically acclaimed by some of the best mathematicians and statisticians of the West.

If one takes a look at the leading European universities, the teaching and research at the University of Pisa (Italy's oldest university) started in 1343. In seventeenth–eighteenth-century France, the two centres of excellence in science education were associated with such mathematical greats as Pierre Simon de Laplace, Simèon Denis Poisson and Garpard Monge. In UK, the first chair in geometry was set up at the Oxford University in 1619, the Mathematics Chair in Abardeen in 1625 and the Lucasian Chair in Mathematics at the University of Cambridge in 1662. In all these Universities, financial and administrative support poured in from the rulers of the country, the local kings and the religious organizations. In contrast, the case of the University of Calcutta and the Indian Statistical University (ISI) was just the opposite. Sir Asutosh in the case of the University of Calcutta and Prof. P. C. Mahalanobis in the case of ISI had to run from pillar to post for funds, administrative help, etc. As India was under foreign domination, the rulers had not the least interest in encouraging study of mathematics and science. This book is a brief real-life account of the unimaginable difficulties, the insurmountable hurdles and the endless struggles which the two great sons of Bengal overcame with their indomitable courage, hard work, dedication and relentless perseverance.

The University of Calcutta, under the leadership of Sir Asutosh Mookerjee, earned a unique distinction of being the first university in India where post-graduate teaching and research in mathematics were initiated. Indian mathematicians from Bengal (presently West Bengal) who carried out pioneering research work in the nineteenth century and early part of the twentieth century had, except for a few, strong connections with the University of Calcutta. An internationally reputed centre for statistical studies and research, Indian Statistical Institute (ISI), Kolkata, also started in Calcutta established and nurtured by Prof. Prasanta Chandra Mahalanobis (P. C. Mahalanobis). The first Indian Institute of Technology (IIT) was also established at Kharagpur in West Bengal. Two brilliant applied mathematicians Prof. B. R. Seth and Prof. A. S. Gupta served there as Faculty Members and made seminal contributions in the areas of solid and fluid mechanics. These are important branches of applied mathematics. These brilliant minds responsible for enriching the mathematical and statistical research climate of Bengal as well as India deserve to be brought to the fore.

In this book, we have discussed the following 16 famous mathematicians and statisticians. In a separate chapter, some brief discussions have been made on the seven remarkable mathematical scientists who worked and enriched ISI with their outstanding contributions.

1. Sir Asutosh Mookerjee (1864–1924)
2. Prof. Syamadas Mukhopadhyay (1866–1937)
3. Prof. Ganesh Prasad (1876–1935)
4. Dr. Bibhuti Bhusan Datta [B. B. Datta] (1888–1958)
5. Prof. P. C. Mahalanobis (1893–1972)
6. Prof. Nikhil Ranjan Sen (1894–1963)
7. Prof. Suddhodan Ghosh (1896–1976)
8. Prof. Rabindra Nath Sen (1896–1974)
9. Prof. Bibhuti Bhusan Sen (1898–1976)
10. Prof. Raj Chandra Bose (1901–1987)
11. Prof. Bhoj Raj Seth (1907–1979)
12. Prof. Subodh Kumar Chakrabarty (1909–1987)
13. Prof. Manindra Chandra Chaki (1913–2007)
14. Prof. Calyampudi Radhakrishna Rao (born 1920)
15. Prof. Anadi Sankar Gupta [A. S. Gupta] (1932–2012)
16. Prof. (Mrs.) Jyoti Das (1937–2015)
17. Some outstanding minds for ISI, Kolkata

This book also discusses a brief historical prelude followed by an introduction to the mathematical scientists mentioned above. This is followed by 16 chapters arranged chronologically and dedicated to the 16 mathematical scientists mentioned above. This book also discusses in brief some exceptionally brilliant statisticians and mathematicians who carried out research in the ISI Kolkata (in Chap. 18). The list includes such eminent personalities as S. N. Roy, D. Basu, J. K. Ghosh, S. R. S. Varadhan, V. S. Varadarajan, K. R. Parthsarathy and R. Ranga Rao. Unfortunately, most of them left Bengal after a couple of years; so long discussions on them have not been carried out. The epilogue has two parts: (i) concluding remarks and (ii) the bibliographies of the scientists mentioned. In case of Prof. R. C. Bose and Prof. C. R. Rao, there are cut-out years, because they left India and settled down in the USA permanently.

Some may question why some well-known mathematical scientists who did outstanding research in the specified time frame have not been included. The answer is as follows.

The two most prominent names of S. N. Bose and M. N. Saha who were both students of applied mathematics and contributed handsomely in the areas of mathematical physics have not been included because their contributions have been discussed in great details in an earlier book by me.[1] So, it was considered prudent not to repeat them in another book by me.

A few good mathematicians were not included also because though they did good research and were good teachers, too, they did not make any pioneering research contributions in their respective fields of research. They also did not inspire and supervise research scholars of high quality.

This book is the first of its kind which gives a well-researched, scientifically accurate historical account of all the above-mentioned facts in a concise and compact form. Naturally, it should be read widely in India and abroad to know and appreciate the glorious achievements that emanated from the state of undivided unified Bengal.

Kolkata, India   Purabi Mukherji

---

[1] Puabi Mukherji and Atri Mukhopadhyay (2018). *History of the Calcutta School of Physical Sciences*. Springer Nature.

Acknowledgements

I wish to acknowledge and convey my grateful thanks to the Indian National Science Academy (INSA), New Delhi, and to the editorial board of the INSA's journal, *Indian Journal of History of Science*, for permitting me to use certain parts of the research papers[2] and materials from the report of the project titled 'Pioneer mathematicians and their role in Calcutta University'[3], all published in the *Indian Journal of History of Science.*

I wish to convey my sincere thanks to Prof. S. Ponnusamy, President, Ramanujan Mathematical Society and the Editor-in-Chief of the *Mathematics Newsletter*, a journal of the said society for allowing me to include my article[4] on Prof. R. C. Bose published in the said journal in this book.

I convey my sincere thanks to Mr. Tapas Basu of the Reprography and Photography Unit of the Indian Statistical Institute, Kolkata, and Prof. Pradip Kumar Majumdar, former Professor of Rabindra Bharati University, Kolkata, for their kind help in sending me the photographs.

I offer my sincere thanks to my friend Dr. Mala Bhattacharjee and my former student Miss Parama Paul for helping me in typing some portions of the manuscript. I want to specially thank my former student Dr. Debashree Ray of Johns Hopkins University, USA, for kindly sending me important materials related to Prof. C. R. Rao.

I am indebted to Prof. Bimal Kumar Roy, former Director of the Indian Statistical Institute, Kolkata, and presently Professor at the Indian Statistical Institute, Kolkata, and Chairman, National Statistical Commission, Government of India, for kindly writing the foreword to this book, in spite of his extremely hectic schedule.

Lastly, I am indebted forever to my beloved father late Sushil Kumar Banerji who was himself a former student of the Department of Applied Mathematics, the University of Calcutta, and also a scholar under late Prof. P. C. Mahalanobis of the

---

[2] Purabi Mukherji, et al.; 46(4), 2011, 48 (2), 2013.

[3] Purabi Mukherji et al., 49 (1), 2014.

[4] Vol. 28 # 4 June–September (2017).

Indian Statistical Institute, Kolkata. My father told me a lot of interesting stories about many of the great personalities discussed in this book and thus indirectly inspired me to write this slim volume.

Purabi Mukherji

Contents

About the Author

Purabi Mukherji has been Counsellor in Mathematics at the Indira Gandhi National Open University (IGNOU), Regional Centre, Kolkata, India, since 1994. Earlier, she served in the Department of Mathematics at the Gokhale Memorial Girls' College, Kolkata, from 1994 to 2014. Dr. Mukherji earned her Ph.D. in applied mathematics from Jadavpur University, Kolkata, in 1987. She received two national 'Best Paper Awards' for her work on 'Mathematical Modelling' in geophysics, awarded by the Indian Society for Earthquake Technology of the University of Roorkee. Since 2010, she has been pursuing research in the field of history of science and has successfully completed two projects funded by the Indian National Science Academy (INSA), New Delhi.

She has published more than 40 research papers in reputed national and international journals and more than 20 scientific articles in Bengali popular scientific magazines. She authored a book titled *History of the Calcutta School of Physical Sciences* (Springer Nature, 2018), *Research Schools on Number Theory in India: During the 20th Century* (Springer Nature, 2020–2021), and *Pioneer Mathematicians of Calcutta University* (University of Calcutta, 2014). Dr. Mukherji is a life Member of the Indian Science Congress Association, Calcutta Mathematical Society, Indian Society of History of Mathematics, Indian Society of Exploration Geophysicists, Indian Society of Nonlinear Analysts, and many others. She is also on the editorial board of the journal *Indian Science Cruiser* being published by the Institute of Science, Education and Culture, Kolkata.

Chapter 1
Historical Prelude and Introduction

1.1 Historical Prelude

The development of mathematical research in India had its own autonomous course. The well-known contributions in subjects like Arithmetic, Algebra, Geometry and Astronomy by the ancient Indians need no recounting. The world of Mathematics has paid merited homage to the astounding discoveries of Zero, the invention of the Decimal System and the initiation of Negative Quantities.

At the initial stage of the British rule, Calcutta (present-day Kolkata) was the capital of India. The colonial rulers governed the rest of the country from this city till 1911. So Bengal was exposed to and influenced by the cultural and educational patterns of the foreign rulers. For historical records, a quote from Sir Charles Wood's Despatch of 1854 is an eye-opener about the intentions of the British rulers. He wrote to Lord Dalhousie: 'We shall govern India for many years but it is clear to my mind that we shall always govern it as aliens.' In fact, Lord Ellenborough warned Wood, 'Education will be fatal to British rule'.

The first three Universities of Bombay, Calcutta and Madras were established by the colonial masters in 1857. The aim of the foreign rulers was certainly not to encourage intellectual activity, but to create a sizable group of English-knowing Indian 'civil servants', who would help them in running the country smoothly. But as already mentioned the British policymakers were still not happy. Wood thought 'Higher Education was at the same time dangerous'. He did not have such misgivings about Bombay and Madras but he was apprehensive about Calcutta.

The University of Calcutta was established on the 24 January 1857 and was used only for giving affiliation to colleges, holding of examinations and conferring degrees.

But fortunately there were exceptional men like Dr. Mahendralal Sircar, Sir Asutosh Mookerjee and Prof. Prasanta Chandra Mahalanobis in Bengal.

The pioneer in the movement for modern Science and Mathematics education in India was Raja Rammohun Roy (1772–1833). In a historic letter written on the 11

December 1823 to Lord Amherst, he requested the colonial rulers to 'promote mathematics, natural philosophy (physics), chemistry, anatomy and other useful sciences.' The letter was completely ignored by the ruling British Government. However the cause was not lost. The 'Science Movement' initiated by Raja Rammohun Roy was carried forward by Derozio (1809–1831), Ishwar Chandra Vidyasagar (1820–1891), Mahendra Lal Sircar (1833–1903) and Sir Asutosh Mookerjee (1864–1924).

Overcoming formidable hurdles and with least help from the colonial rulers, Dr. Mahendra Lal Sircar in 1876 established the 'Indian Association for the Cultivation of Science' (IACS) in Calcutta, the then Capital of British ruled India. It was founded with money collected from public donations (both large and small amounts). Right at the outset Dr. Sircar had declared that the newly established Institute would combine character, the scope and objectives of the 'Royal Institute of London' and the 'British Association for the Advancement of Science'. He also took two vital and very significant policy decisions; namely, that the institution will impart scientific knowledge to the masses and it will be managed and controlled solely by indigenous Indians. Dr. Sircar always felt that the main reason for the lack of development of India was due to backwardness of the Indians in science. He was confident that given adequate and necessary opportunities, Indians had the potential to master modern science. The establishment of IACS in Calcutta was the first major step of the 'Indian Science Movement'. Sir Asutosh Mookerjee following the lead of Dr. Sircar was among the first to propose the idea that a University should be a centre of independent intellectual activity of research and teaching of high quality at all levels. But to achieve that Sir Asutosh had to face a gigantic task. The Indian Universities Act of 1904 was a very controversial act right from the beginning. As the Sadler Commission Report remarked, 'The Indian universities under the new Act were the most completely governmental universities in the world'.

1.2   Sir Asutosh Mookerjee's Role in Creating the Post-graduate Departments of Mathematics in Calcutta University

The Governor General and Viceroy of India, Lord G. N. Curzon (1859–1925), in 1902 had appointed a commission comprising only Europeans. As the commission meant for reform of Indian Universities did not include a single Indian, for obvious reasons the public sentiment was strongly against it. The British Government, in order to pacify the raging public sentiments, made Sir Gooroodas Banerjee (who was the first Indian Vice-Chancellor of Calcutta University) a Member of the commission. Asutosh Mookerjee was taken in as a co-opted Member to act as a provincial representative educationist.

As far as teaching functions were concerned the 1904 Act was merely a permissive one. Asutosh Mookerjee believed that a combination of research and teaching was the inalienable basic principle of any University. Since London University, which was the role model of Indian Universities, was undertaking teaching functions, it was felt that Indian Universities should also do it within defined limits.

The main provisions of the 'Universities Act' of 1904 had been summarized by eminent educationist G. K. Gokhale as follows:

The Universities Act of 1904 dealt with the expansion of the functions of the University. Secondly it dealt with the constitution and control of the Universities, and thirdly it dealt with the control of the affiliated colleges.

Indian educated class including G. K. Gokhale was not at all hopeful about the first objective, and they apprehended that it will remain 'an aspiration, a remote ideal'.

For general interest, that momentous clause was:

The University shall be and shall be deemed to have incorporated for the purpose (among others) of making provisions for the instruction of students, with power to appoint University Professors and Lecturers, to hold and manage educational endowments, to erect, equip and maintain University libraries, laboratories and museums, to make regulations relating to the residence and conduct of students, and to do all acts, consistent with the Act of Incorporation and this Act, which tend to the promotion of study and research.

But Asutosh Mookerjee with his iron wills, his indomitable courage and perseverance and his enormous self-confidence and dynamic and diplomatic personality seized on that one section of the Universities Act of 1904.

Going back a little in time, it may be noted that the proposal to appoint Asutosh Mookerjee as Vice-Chancellor of Calcutta University was initiated by the then Home Secretary H. H. Risley. In a long note he wrote: 'Sir A. Pedler will leave India about the end of March. It is necessary to consider the question of appointing his successor.

I have no hesitation in saying that the Hon'ble Mr. Justice Mookerjee is marked out by his scientific attainments, his long connection with the University and the work he has done for it and by his official position as conspicuously qualified for the post of a Vice-Chancellor. Dr. Mookerjee was appointed a Fellow of the University in January, 1889 and has been a Member of the Syndicate for more than 16 years. For the last eleven years he has been President of the Board of Studies in Mathematics and he has conducted the highest examinations in Mathematics and law since 1887. Dr. Mookerjee had a brilliant academic career, his chief distinctions being the Premchand Roychand Studentship, the degree of Doctor of Law and the Tagore Law Professorship. In his special subject, Pure Mathematics, he has a European reputation and the results of his original researches have been embodied with his name in standard Cambridge text books.

The appointment of a distinguished Indian as Vice-Chancellor would undoubtedly be popular and would tend in some degree to discourage the idea that the sole purpose of the Universities Act was to tighten official control over the Universities. Finally I may mention that it is an advantage for the Vice-Chancellor to be a judge of the High Court, since the political faction in the Senate is composed mainly of pleaders and they are amenable to a judge before whom they have to appear in Court than to an executive official.'

Asutosh Mookerjee was appointed as the Vice-Chancellor of the Calcutta University on the 31 March 1906 for a period of two years. When he first assumed charge, the then Viceroy of India Lord G. E. Minto (1751–1814) had advised him 'to work in concurrence with the Senate in such manner as might appear to my judgement to be in the best interests of the University.'

On the basis of his superlative performance Asutosh Mookerjee was reappointed three times, and so he continued in office from the 31 March 1906 to the 30 March 1914. All through this period he kept in mind the sage advice of Lord Minto.

After assuming office, Asutosh Mookerjee seized on that 'permissive clause' pertaining to teaching and research. He used it as a magic wand to convert the University of Calcutta into a great centre for teaching and research. He had been long waiting for such an opening and was quick to take advantage of it. At that time because of the political changes all around the world, 'Asian Nationalism' was gradually awakening and India was no exception. In 1905, Lord Curzon's decision to divide Bengal gave rise to tremendous popular resentment. Bengal and practically the whole of India were in open rebellion against the British rule. Asutosh Mookerjee was a child of the times. He worked relentlessly to bring about an intellectual regeneration and a nation-wide progress in education. This was his chosen tool. His lifelong mission was to raise the motherland to the glorious pedestal of intellectual supremacy.

Asutosh Mookerjee himself had obtained his M.A. in Mathematics from the prestigious Presidency College. In his famous 'Diary' he had praised three of his Mathematics teachers profusely. They were Professors W. Booth, J. McCann and J. A. Martin. But Asutosh Mookerjee was not happy with some of the other Faculty Members. He always felt Post-Graduate teaching need not aim at fostering the spirit of original research but it should somehow be inspiring so that specially endowed youthful minds may get a chance to catch sparks of its fire.

Within one year of assuming office as Vice-Chancellor, Asutosh Mookerjee set about the Herculean task of establishing a teaching University. His challenging task may roughly be classified as follows:

(1) He tried to arrange for instructions to be given to the Post-Graduate students, through University Lecturers. They were mainly recruited from the Faculty of affiliated colleges and also drawn from well-known scholars of distinction.
(2) He tried to establish University chairs which were filled up by eminent academicians.
(3) He appointed regular fulltime University Professors, Readers and Lecturers. With huge monetary help received from Tarak Nath Palit and Rashbehary Ghosh, in 1914 Asutosh Mookerjee laid the foundation stone of the University College of Science. This was the unveiling of the new horizons aimed at enriching Science and Mathematics education in the country.
(4) In the final stage, Post-Graduate study and research in Bengal was centralized in the hands of the University and the Post-graduate Departments of Arts and Science were established.

But the task that Asutosh Mookerjee set for himself was not at all easy for various reasons. The first and foremost difficulty was related to funds. The colonial rulers were not at all willing to allot funds to the University for organizing higher education and research. The Vice-Chancellor was forced to use the surplus funds of examination fees to carry out any constructive measure. It was a very unhealthy state of things. Even in such a situation in a letter Carmichael wrote to Lord Hardinge: 'The financial condition of the University must be fully investigated, not with reference to its future needs or the making of future grants but with a view to finding out the utility— (what we have never obtained) a clear but exhaustive explanation of its position.' This clearly reflects the attitude of the British Government.

Another incident may also be recalled. Once Lord G. N. Curzon, the then Viceroy of India, had taunted G. K. Gokhale and Asutosh Mookerjee and remarked that the rich men of India did not readily make endowments to Indian Universities. He had said that there had been no endowments during the previous forty years. In a sense it was true in the case of Calcutta University, as Indian aristocracy had not imitated the munificence of Premchand Roychand and Prasanna Kumar Tagore. But Asutosh Mookerjee's whole-hearted devotion and his commitment as a Vice-Chancellor dedicated to the welfare of the University inspired confidence in the moneyed class after a prolonged period of distrust. The Maharaja of Darbhanga, eminent legal luminaries Sir Tarak Nath Palit and Sir Rashbehary Ghose came forward with huge donations. That helped Vice-Chancellor Asutosh Mookerjee to initiate his dream projects.

In 1906 Asutosh Mookerjee took charge as Vice-Chancellor of Calcutta University. The then Viceroy of India, Lord G. E. Minto (1851–1914), who was also the Chancellor of Calcutta University and the Home Secretary Mr. H. H. Risley were both favourably disposed towards him. In fact they admired him. In his monumental task of reforms and reorganizations of the educational system he had assurances of sympathy and support from the Government of India. In his last Convocation address as the Chancellor of Calcutta University, on the 12 March 1910, paying rich tribute to Asutosh Mookerjee, Lord Minto said: 'Now that my high office is drawing to a close I rejoice to feel that that the administration of this great University will continue to benefit from your distinguished ability and fearless courage.'

But after the departure of Lord Minto, the next Viceroy Lord Hardinge (1858–1944) was indifferent towards Asutosh Mookerjee. As a consequence, sympathy and support were replaced by opposition and hostility. A period of controversy between the Government of India and the University of Calcutta ensued. There was unending interference in the matters related to the University from the side of both the Central and the Bengal Governments. But Asutosh Mookerjee had in him a rare combination of wide and deep intellectual interests and outstanding administrative ability and statesmanship.

As rightly pointed out by Prof. R. N. Sen, a famous Hardinge Professor of Higher Mathematics: 'In spite of many handicaps inherent in the Act of 1904 and against heavy odds, he carried the Senate with him to make plans, schemes and regulations for stimulating and spreading education in the country .... His crowning achievement was the creation in 1917 of the Post-Graduate Departments of Teachings in Arts and Science in the University which provided opportunity and incentive for higher studies and research. To achieve all these objectives he was able to raise generous and princely donations and to attract the most learned and talented persons in science, humanities and letters from all over India to run the Post-Graduate Departments. The University was transformed from being merely an affiliating and examining institution into an organization with the added responsibility of disseminating and unfolding knowledge with the motto 'Advancement of Learning.'

Thus overcoming tremendous difficulties, crossing monumental hurdles, finally Asutosh Mookerjee was able to fulfil his dreams of creating the two Departments of Mathematics in the University of Calcutta.

He slowly started to recruit exceptionally talented mathematicians to create new departments of Mathematics. In 1907, C. Little of Presidency College was appointed as a University Lecturer in Mathematics. In 1909, the well-known mathematician and a Professor of Mathematics from Presidency College, C. E. Cullis was appointed as University Reader in Mathematics. In 1911, Syamadas Mukhopadhyay, an eminent mathematical researcher and also a Professor of Presidency College was appointed as a University Lecturer in Mathematics. In 1911, the chair of 'Hardinge Professor of Higher Mathematics' was created in the Department of Pure Mathematics. The first appointee to the Hardinge Professorship was a famous British mathematician Prof. W. H. Young FRS. Prof. Young was earlier a Professor of 'Higher Analysis' in the University of Liverpool and a Fellow of Peter House College, Cambridge. He spent about three years serving the Department of Pure Mathematics. In 1913, Sir Rashbehary Ghosh, an eminent jurist and erudite scholar made a generous gift of money to the University of Calcutta for the promotion of scientific and technical education and for the advancement of research in pure and applied sciences. The prescribed condition in his endowment was to create four University Professorships in different scientific disciplines. Accordingly the 'Rashbehary Ghose Professorship' in Applied Mathematics was created. The newly created Mathematics Department was bifurcated into two separate departments of 'Pure Mathematics' and 'Mixed (later named as Applied) Mathematics'. The first holder of the chair of the Rashbehary Ghose Professorship was an eminent Indian mathematician Dr. Ganesh Prasad.

The Post-Graduate Department of Pure Mathematics started functioning from 1912. As already mentioned Prof. Young was the 'Hardinge Professor' and Head of the Department of Pure Mathematics. The four Lecturers in the Department were Dr. Syamadas Mukhopadhyay, Haridas Bagchi, Indubhusan Brahmachari and Phanindranath Ganguly. Apart from these fulltime Faculty Members, two other eminent mathematicians who were Readers in the University delivered lectures on specified topics to Post-Graduate and advanced level students and initiating them to the world of original research in Mathematical Sciences.

The Post-Graduate Department of Mixed (Applied) Mathematics started functioning from 1914 under the stewardship of Prof. Ganesh Prasad. He was the 'Rashbehary Ghose Professor' and Head of the Department of Mixed (Applied) Mathematics. Initially he was assisted by two of his research scholars, Dr. Sudhangshu Kumar Banerjee and Dr. Bibhuti Bhusan Datta.

The Post-Graduate Departments centralized higher study and research in Mathematics. This was a step of far-reaching significance. Both the two Departments produced brilliant mathematicians who made notable contributions to Mathematical Sciences in future. Many of them have been included in this monograph. They carried on the legacy of research and teaching which Sir Asutosh Mookerjee had so carefully cultivated, nurtured and thus initiated the spirit of modern Western Mathematics not only in Bengal but in whole of India.

In 1914, while laying the foundation stone of the new University College of Science in Rajabazar area of Calcutta, he proudly announced: 'I trust, I may be permitted to dwell without impropriety on the gratifying circumstance that of the six Professors, fully one half come from provinces other than Bengal. We are proud indeed, to have in our teaching body those distinguished representatives of Madras, Bombay and the United Provinces. No stronger testimony is needed to emphasize the cosmopolitan character of science, and I fervently hope that although the College of Science is an integral component part of the University of Calcutta, it will be regarded not as a provincial but as an All-India College of Science to which students will flock from every corner of the Indian empire, attracted by the excellence of the instruction imparted and the facilities provided for research.'

Thus he served his motherland and was an important player in furtherance of the true progress of his countrymen by diverting some of the best among them from the chief industries of the land, law and government service, to scholastic careers.

1.3 Professor P. C. Mahalanobis's Role in the Establishment of the Indian Statistical Institute (ISI), Calcutta and Starting the Post-graduate Course in Statistics in the University of Calcutta

After completing his education at King's College, Cambridge, England P. C. Mahalanobis returned to Calcutta. In 1915, he joined the 'Indian Educational Service' (IES) and started working as a Lecturer in Physics in the renowned Presidency College, Calcutta. In 1922 he became Professor of Physics and continued serving in the Department of Physics, Presidency College. Though he was trained as a physicist in Cambridge, but a few months before his departure from there, P. C. Mahalanobis developed a keen interest in both pure and applied Statistics. His exposure to the work of famous British statisticians, which he picked up from the articles published in 'Biometrica', helped him to become an independent researcher in the discipline. Along with his work as a Professor of Physics, he had already gone through more than a decade of long preparatory work, carrying out statistical studies. These studies comprised a number of innovative statistical analyses of a variety of data, anthropometrical, meteorological and those related to agricultural experiments.

In 1924, Prof. Mahalanobis had managed to get a small anteroom in the Baker laboratory of Presidency College, where he set up what was then called a 'Statistical Laboratory'.

The establishment of the 'Indian Statistical Institute' (ISI) in that small room of Presidency College, the history of its phenomenal development can be attributed only to the individual efforts and imaginative planning of Prof. P. C. Mahalanobis.

On the 14 December 1931 in a meeting convened by Professor P. C. Mahalanobis, Professor N. R. Sen, Rashbehari Ghosh Professor of the Department of Applied Mathematics and Professor Pramatha Nath Banerjee, Minto Professor of the Department of Economics, both from the University of Calcutta, it was unanimously resolved that an Indian Statistical Institute (ISI) be started and that the noted industrialist Sir R. N. Mookerjee be requested to accept the office of President of the Institute. Following this, the Institute was formally registered on the 28th April, 1932.

After the registration was complete the Institute (ISI) started functioning from the tiny 'Statistical Laboratory'. In fact both started operating as a single operating unit. To start with Prof. Mahalanobis had only one close associate and collaborator. He was a brilliant young man named S. S. Bose. He worked with immense sincerity and dedication till his untimely demise in 1938. Shortly after that J. M. Sengupta joined ISI. He did all types of work as and when necessary. Gradually he developed an excellent feel for Statistics. The transition from the 'Statistical Laboratory' to 'Indian Statistical Institute' still continued to be a one-man show. But the great visionary Prof. Mahalanobis initiated a whole range of activities over the next two or three years. The statistical analysis of data by him during the previous decade that had earned him recognition still continued in full force. The ISI was continuously engaged in agricultural experiments using techniques discovered by Prof. Mahalanobis and were almost identical to those introduced by the world-renowned statistician Prof. R. A. Fisher in England. In fact it is on account of the statistical work on the discipline of agriculture that ISI obtained an annual grant of Rs. 2500/- for 3 years from the 'Imperial Council of Agricultural Research' (ICAR), Government of India. At that point of time, this was the only assured income of ISI.

Prof. Mahalanobis had a deep concern regarding the correctness of the collected data. Right from the beginning of his now famous 'Sample Survey Projects', he had taken great care for minimizing sampling errors. In 1936, he derived a technique called 'interpenetrating subsamples'. In this method two or more independent groups of independent investigators were required to take measurements of two subsamples of the same sample under the constraint that they would not meet. This was novel methodology having the advantage of assessing sampling error, as well as controlling other types of errors. These additional precautionary exercises naturally pushed up the expenses of conducting survey work. But Prof. Mahalanobis convincingly argued that the benefits were much greater and they outweighed the marginal increase in expenditure.

After a couple of years ICAR doubled the annual grant and made it Rs. 5000/- per annum. The earnings from the projects had to be spent very judiciously, because savings from these were absolutely necessary. The ISI was gradually expanding and the financial commitments had to be met. The monthly salary of the workers was an important issue. Incidentally, it is interesting to note that the compulsion of cost-cutting gave impetus to another type of statistical research, namely field studies on various optimization problems associated with sampling. In those days they were called 'statistical experiments'.

In 1933, Prof. Mahalanobis started the publication of 'Sankhyā', the first statistical journal of India. This venture not only required money, but the physical efforts involved was really daunting. A small hand operated type-casting machine was imported and locally made hand-cut dice were used for the mathematical symbols. As the frequency of publications increased better machines became necessary. It was an intensely difficult venture. In retrospect, even Prof. Mahalanobis himself had commented: '… an adventure, even foolhardie, to have started a statistical journal in India 30 years ago when our resources in research and material equipment were meagre.' But like most of Prof. Mahalanobis's other bold initiatives, 'Sankhyā' too succeeded as an international Journal of Statistics. From the standpoint of ISI, it was of invaluable help, because it brought to the fore the research and other statistical activities conducted at the Institute.

For ensuring regular flow of funds for uninterrupted publication of 'Sankhyā', in 1936, ISI had to make an agreement with the State of Holkars (a princely state in pre-independent India). The deal was for getting an annual grant of Rs. 1000/- earmarked for 'Sankhyā', ISI agreed to prepare and publish in 'Sankhyā' an annual statistical review of the Holkar State on the basis of materials received from the State.

With research materials produced by the enthusiastic researchers like J. M. Sengupta, R. C. Bose, S. N. Roy, Prof. Mahalanobis himself and others along with the reports of the 'sample surveys' conducted related to 'flood control studies in North Bengal', in the 'Damodar Valley' and in the 'Mahanadi Valley', the 'Bengal Jute Survey', the 'Bihar Crop-survey', etc., 'Sankhyā' had a smooth sailing. Even today it still functions as a medium of advanced statistical publications in India.

It is well known that P. C. Mahalanobis was extremely close to the Nobel Laureate poet Rabindra Nath Tagore. On a request from Professor Mahalanobis, Tagore wrote a beautiful couplet in the second volume of 'Sankhyā'. It is as follows: 'These are the dance steps of numbers in the arena of time and space, which weave the maya of appearance, the incessant flow of changes that ever is and is not.'

In the thirties of the last century, Prof. Mahalanobis was trying hard to achieve a dual goal; recognition of his brain-child the ISI and the recognition for the discipline of Statistics as well. At that point of time, in India, in the academic circles as well as in the echelons of power very few people had a fair understanding about what Statistics was and why were it necessary. Prof. Mahalanobis left no stone unturned to attain his targets. On the academic front, he first tried to lobby with the authorities of the 'Indian Science Congress' to open a separate section for Statistics. His attempts were not successful. But he was a man who never gave up. So he decided to hold 'Indian Statistical Congress' every year after the annual Science Congress ended in the very same venue. The first such Congress was held in December–January, 1937 under the Chairmanship of the world-renowned statistician Prof. R. A. Fisher in Calcutta. After several such Statistical Congresses, the 'Indian Science Congress' decided to include Statistics under the banner of the Mathematics Section from 1942. From 1945, a separate section for Statistics was formed. Meanwhile during Prof. Fisher's visit to Calcutta, through his good offices, some senior British civil servants of India started taking note of the work that was being done by Prof. Mahalanobis. Finally, on the 15 December 1937 the Viceroy Lord Linlithgow officially visited the Indian Statistical Institute. So the doors of administrative awareness about the ISI opened up. The academic community was gradually developing an interest in the activities of the ISI as well as the discipline of Statistics.

But in spite all the important survey work and theoretical research done by the group led by Prof. Mahalanobis, the financial difficulties persisted, as there was no provision for regular funding to ISI.

In pre-independent India, Professor P. C. Mahalanobis had to run from pillar to post to keep the Institute going. Literally ISI lived from hand to mouth in those times, the basic source of finance continuing to be ad hoc project grants made by Central Government bodies like ICAR and the provincial Bengal government. Another problem was related to the space. As Professor Mahalanobis had retired from the Presidency College, the Government of Bengal was putting pressure on him to vacate the space in Presidency College. The present building that houses the ISI at B. T. Road was not built till then.

When C. D. Deshmukh became the Governor of the Reserve Bank of India in 1945, things started improving. After India gained her independence, the first Prime Minister of India, Jawaharlal Nehru took a personal interest in the matters related to ISI. He even deputed his one-time private secretary Mr. Pitambar Pant to go to ISI and take courses on Statistics.

Finally after many years of struggle, in 1960, Professor Mahalanobis was successful in obtaining for the Institute a guarantee of support from the Central Government. With positive initiative and help from India's first Prime Minister Jawaharlal Nehru, bills were passed in both houses of the parliament in 1959, and the Indian Statistical Institute Act finally became operational from April, 1960. By virtue of the act, the Indian Statistical Institute was recognized as an 'Institute of National Importance'.

This was the end of Professor Prasanta Chandra Mahalanobis's long, lonesome and strenuous struggle for the fulfilment of his dream.

Another major contribution of Professor P. C. Mahalanobis as an institution builder relates to his role in persuading the authorities of the Calcutta University to open a Post-Graduate Department dedicated to Statistics. As no institutes or University offered a formal course on Statistics per se, Prof. Mahalanobis strongly felt the need for such facilities.

There is a funny anecdote related to this event. When Professor Mahalanobis first broached the idea of opening the Post-Graduate Department in Statistics, many Council Members of Calcutta University apprehended that there were not enough reading matters available in Statistics, hence it would not be wise to open the said department. To counter the argument, the story goes that Professor Mahalanobis employed some hired porters to carry three basketfuls of Biometrica volumes and other treatises to the venue of the Calcutta University Council meeting.

Jokes apart, Professor P. C. Mahalanobis with his unmatched power of persuasion, patience and perseverance was finally successful in convincing the Calcutta University authorities and the new Post-Graduate Department of Statistics started to function from July, 1941. This was the first centre of Post-Graduate studies in Statistics in India. In fact, at that point of time, not only in India, but in the whole world there were very few departments dedicated to the teaching of Statistics as a separate subject.

According to C. R. Rao, who was a student in the first batch, 'The courses taught were of reasonably good standard and properly balanced in the theory and applications of Statistics. The curriculum of courses in Statistics developed in the early 40's in the Calcutta University was adopted by the other Universities in India which later introduced the Master's program in Statistics'.

Professor Mahalanobis was also the leading spirit in creating the under graduate department of Statistics in the Presidency College, Calcutta. Being a Professor of the same college, he could play a more pro-active role there.

Thus Calcutta became a great centre for teaching and research in Statistics along with the unique 'sample survey' projects spearheaded by Prof. P. C. Mahalanobis. The projects fetched money from the sponsors, which made scientific activities of the Indian Statistical Institute possible. These activities also strengthened the credibility of both Prof. Mahalanobis and his dream creation, the ISI.

1.4 Introduction

A brief sketch about the scientists already mentioned in the 'Preface' is given below.

1. Sir Asutosh Mookerjee was an extremely talented multifaceted versatile man. He almost single-handedly started the culture of mathematical research in India. He had no one to guide him. He carried out research on his own and made some original contributions mainly in the areas of Geometry, Differential Equations and Hydrokinetics. The total number of his personal research papers is 17, and they were published in different national and international journals. Some of his research papers aroused a lot of interest among the British school of mathematicians of those times.

2. Prof. Syamadas Mukhopadhyay was the first Indian to obtain a doctorate degree in Mathematics in India. Inspired from his student days by his favourite teacher Prof. W. Booth of Presidency College, Calcutta and later by Sir Asutosh Mookerjee, after joining the Pure Mathematics Department of Calcutta University, he seriously carried out research in various branches of Geometry and made notable contributions in various branches of the discipline. He is internationally acclaimed for what is now known as Mukhopadhyay's 'Four Vertex Theorem'. He also did some fundamental research in the field of plane Hyperbolic Geometry and Differential Geometry of curves. He published 30 original research papers in national and international journals of repute. As a research guide he produced two brilliant mathematicians namely Prof. R. C. Bose and Prof. R. N. Sen. They made remarkable contributions and will be discussed later in this write-up.

3. Prof. Ganesh Prasad often referred to as the 'Father of Mathematical research in India' served the Calcutta University in two phases. He acted as the Head of the Department and held prestigious 'Chair Professorships' in both the Departments

of Pure as well as Mixed (later known as Applied) Mathematics. After earning M. A. degree from the Universities of Allahabad and Calcutta, he went abroad on a Government of India scholarship. He pursued higher studies at the Universities of Cambridge in England and subsequently at the Göttingen in Germany. He was a student of the famous German mathematician Felix Klein. Prof. Prasad made notable research contributions in Theory of Potentials, Theory of Functions of a Real Variable, Fourier Series and Theory of Surfaces. He wrote and published about 50 original research papers in different national and international journals. He also authored a number of books on different mathematical topics. Some of them are still considered as classics. But apart from his own achievements, he inspired young students to take up mathematical research seriously and produced a number of famous mathematicians in the process.

4. Next a remarkable mathematician Dr. Bibhuti Bhusan Datta [B. B. Datta] has been discussed. In the backdrop of the early twentieth-century mathematical scenario of this country, Dr. Datta's research contributions in 'Fluid Mechanics' are certainly rich, but his contributions in the field of 'History of Mathematics' are stupendous. It was his teacher and colleague Prof. Ganesh Prasad who inspired him. Dr. B. B. Datta took up the challenge and the output was fabulous. He wrote nearly 60 articles on the contributions of the Hindu and Jaina mathematicians. In 1931, Dr. Datta delivered six lectures on ancient Hindu Geometry. In 1932, Calcutta University published them in the form of a book titled 'The Science of Sulba—A study of Hindu Geometry'. In 1935, Dr. Datta in collaboration with Dr. A. N. Singh of Allahabad University published the classical treatise titled 'The History of Hindu Mathematics' (Part I and II). This is considered as an all-time classic in this discipline.

5. Prof. Prasanta Chandra Mahalanobis [P. C. Mahalanobis] FRS was initially trained as a physicist at the University of Cambridge, England. At the end of his stay his tutor drew his attention to the well-known Statistical journal 'Biometrica'. That initiated him to this new area of modern mathematics and he gradually mastered the subject. With his brilliant mind he was able to foresee its applications in the context of India. His personal contributions in Statistics which have been internationally acclaimed comprise 'Mahalanobis Distance' a statistical measure and his pioneering studies in anthropometry in India. He also contributed greatly to the design of large scale 'Sample Surveys' in different areas in India. Apart from his personal contributions, his greatest achievement was in setting up the 'Indian Statistical Institute' in Calcutta. The Institute has over the years produced world-class statisticians and mathematicians and is still considered to be a centre of teaching and research of international standard. Prof. P. C. Mahalanobis is rightly considered as the 'Father of Statistics in India'. His birthday the 29th June is celebrated as the 'National Statistics Day' in India as a tribute to the great scientist.

6. Prof. Nikhil Ranjan Sen [N. R. Sen] was appointed for a teaching position in the newly set up Department of Mixed Mathematics (later renamed as Applied Mathematics) of Calcutta University in 1917 by the illustrious mathematician Vice-Chancellor Sir Asutosh Mookerjee. N. R. Sen was inspired by Sir Asutosh

to take up serious research in the field of Mathematics. In his long and illustrious career he carried out research over a wide range of subjects and topics. He obtained his D.Sc. degree from the University of Calcutta for his original contributions in Applied Mathematics. In the initial years Dr. Sen took a keen interest in the theory of Newtonian Potential and also in the Mathematical theories of Elasticity and Hydrodynamic Waves. Soon after completing his D.Sc. degree, Dr. Sen went to Germany and worked with Prof. Max Von Laue in the General Theory of Relativity and on Cosmological problems. During his stay there he also started research work on Quantum Mechanics. After his return to India, in 1924 N. R. Sen took over as 'Rashbehary Ghosh Professor' of Applied Mathematics. Apart from his own brilliant research contributions, he introduced modern subjects in the Post-Graduate curriculum and inspired his young colleagues and research scholars to take up original and challenging problems and solve them in new areas like Relativity, Astrophysics, Quantum Mechanics, Geophysics, Statistical Mechanics, Fluid Dynamics, Magneto-Hydrodynamics, Elasticity and Ballistics. He was the fountain-head of inspiration to research workers in Calcutta University and under his dynamic leadership the Department of Applied Mathematics became a vibrant centre of teaching and research and earned a reputation throughout India, which would be hard to match. He was very rightly called the 'Father of Applied Mathematics' in India. The total number of scientific publications of Prof. N. R. Sen in famous national and international journals is 46.

7. Prof. Suddodhan Ghosh [S. Ghosh] was a brilliant mathematician and great teacher, but he was a greater human being. In some ways he was unique. He was a man who lived a life of self-denial and his style of living was simple and frugal. But anonymously he donated money to religious, social and charitable organizations. He served the Calcutta University for thirty-five years and was a Faculty Member both in the Departments of Applied and Pure Mathematics at different times. He was fascinated by the teaching of his famous teacher Prof. S. N. Bose FRS and was irresistibly drawn towards Mathematical Theory of Elasticity. He published his first research paper in this area while he was still a student of the Post-Graduate class in the Department of Applied Mathematics. Prof. Ghosh's main research work comprises problems in Fluid Dynamics and Elasticity. He was a leading researcher in the above two fields in India. His important research papers have been referred to in standard text books on Elasticity and relevant books of reference. He published about 30 original research papers. Except for one, all the remaining papers were by published in Indian Journals and 28 of them in the 'Bulletin of the Calcutta Mathematical Society'. In Elasticity, Prof. Ghosh solved problems on 'Bending of Elastic Plates', 'Problems of Dislocation', 'Plane Problems', 'Vibration of Ring' and 'Torsion and Flexure Problem'. In the well-known book titled *'Resistance des Materiaux'* by Robert L'Hermite the problems of 'Torsion and Flexure' as solved by Prof. Ghosh and his student D. N. Mitra have been discussed in details. He was admired and respected by his students and most of them held him in awe.

8. Prof. Rabindranath Sen [R. N. Sen] was a student and later a Faculty Member of the Department of Pure Mathematics in Calcutta University. He earned his Ph.D. degree in a short span of two years working under the supervision of Sir E. T. Whitaker in the Edinburgh University of Scotland. He returned to India in 1930. He joined his alma mater, the Department of Pure Mathematics, Calcutta University as a Lecturer. From 1954, for seven years, he served the same Department as the Head as well as the 'Hardinge Professor' of Pure Mathematics. In his personal research Prof. Sen made outstanding contributions in the fields of Differential Geometry of Riemannian and Finsler spaces. He discussed the connection between Levi–Civita's parallelism and Einstein's tele-parallelism. He also made original research about curvature of hyper space and studied two types of parallel displacements in Riemannian space by the introduction of two arbitrary symmetric linear connections. Prof. Sen's investigation on an arbitrary parallel displacement in a metric space resulted in the discovery of an algebraic system of affine connections in which Levi–Civita parallelism could be identified. This work is supposed to be highly significant and Prof. I. M. H. Etherington of Edinburgh University referred to it as 'Senian Geometry'. Prof. R. N. Sen published 63 research papers in well-known national and international journals.

9. Prof. Bibhuti Bhusan Sen [B. Sen] was a brilliant student of the Department of Applied Mathematics, Calcutta University. He had the good fortune of being taught by such stalwarts as Prof. S. N. Bose FRS and Prof. N. R. Sen. Inspired by Prof. S. N. Bose, young B. Sen took up research in the field of Solid Mechanics seriously. He was the man who practically initiated modern type of research in the field of Solid Mechanics. He made notable contributions in Elasto-Dynamics, Thermo-Elasticity, Visco-Elasticity, Magneto-Elasticity, Piezo-Elasticity, Plasticity, Couple stresses, Theoretical Seismology, etc. He was a superlative research guide and perhaps guided the largest number of Ph.D. students (about 50) among all the contemporary mathematicians of his time. Apart from Elasticity, he also guided some scholars to do research in Fluid Dynamics. He also authored some notable books on Higher Mathematics. His monographs on 'Special Functions' and 'Laplace Transform' were published in his lifetime. They were frequently used by teachers and researchers. His book on 'Numerical Analysis' was published posthumously. Prof. Sen built up a strong school of research in Solid Mechanics in India. A very remarkable fact is that Prof. B. Sen never allowed his own name as a co-author with any of his students in their publications. Prof. B. Sen singly published 57 research papers in famous national and international journals.

10. Prof. Raj Chandra Bose [R. C. Bose] after completing his M.A. from Delhi University came and joined the Pure Mathematics Department, Calcutta University. He became a legendary research worker in Geometry under the guidance and inspiration of Prof. Syamadas Mukhopadhyay of the same Department. Prof. Bose is internationally acclaimed for his contributions in Geometry and Mathematical Statistics. The 'Father of Statistics in India', Prof. P. C. Mahalanobis persuaded R. C. Bose to join the newly set up Indian Statistical Institute

(ISI) in the late thirties of the twentieth century. R. C. Bose was a pioneer in using finite geometries, finite fields and combinatorial methods in the construction of designs. His famous work in collaboration with another brilliant applied mathematician S. N. Roy on 'Distribution of the Studentised D^2—statistic' brought him international recognition from stalwart statisticians of the time. Later in 1941 he joined as a Lecturer in the newly established Department of Statistics of Calcutta University. He however continued to work at the ISI in a part-time capacity. In 1947, R. C. Bose obtained his D.Lit. degree from the University of Calcutta for his original work on Multivariate Analysis and Design of Experiments. In 1949, Dr. R. C. Bose permanently emigrated to USA. and joined as a Professor in the University of North Carolina. While in USA. he continued doing research in diverse areas of Mathematics and Statistics. He made pioneering work in the field of error correcting codes. Prof. R. C. Bose published a total of 144 original research papers in important national and international journals.

11. Bhoj Raj Seth [B. R. Seth] had a uniformly brilliant academic career. In 1927 he obtained the B.A. (Honours) degree in Mathematics and in 1929 he got his M.A. degree in Mathematics from Delhi University. He stood First with a First class in both the examinations. He obtained a prestigious Central Government Scholarship and proceeded to England for higher studies. He specialized in Elasticity, Photo-Elasticity, Fluid Mechanics and Relativity and obtained the M.Sc. degree from the London University in 1932. He then went to Germany and pursued studies at the University of Berlin. Working under the guidance of Prof. L. N. G. Filon, B. R. Seth obtained the Ph.D. degree from the University of London in 1934. The title of his thesis was '*Finite Strain in Elastic Problems*'. In 1937, Dr. B. R. Seth obtained the D.Sc. degree from the University of London. Dr. Seth made original contributions in the fields of Elasticity and Fluid Mechanics. He also carried out some experimental work on photo-elastic properties of celluloid in collaboration with Professor Harris. He also made notable contributions in areas of Plasticity, Boundary Layer Theory, Potential Theory and allied subjects. Prof. Seth's research contributions have been referred to in many standard text books on mathematical Theory of Elasticity. After returning to India in 1937, he served the Hindu College in Delhi for eleven years. Subsequently in 1949, he went to USA and served as a visiting Professor at the Iowa State University for two years. As soon as he returned to India, he was appointed as a Professor and Head of the Department of Mathematics at the newly established first Indian Institute of Technology (IIT) at Kharagpur, West Bengal. He served the Institute for sixteen long years and devoted himself whole heartedly to build up an exemplary Department of Applied Mathematics. Under his leadership a strong school of research flourished in Elasticity, Plasticity, Rheology, Fluid Dynamics and Numerical Analysis. The said Department achieved the status of one of the foremost centres of research in those disciplines in India, as well as beyond the shores of this country. He was a dynamic leader both at home and abroad. He was a great mathematician, research guide and institution builder all combined

into one great personality of his time. He published about 120 original research papers in noted national and international journals.

12. Subodh Kumar Chakrabarty [S. K. Chakrabarty] was a student and later a Faculty Member in the Department of Applied Mathematics, Calcutta University. In 1932 he stood First with a First class in M.Sc. (Applied Mathematics). He won the 'Sir Asutosh Mukherjee Gold Medal' for securing highest marks among all successful candidates in all subjects that year. In 1935 he joined his alma mater as a Faculty Member. He was the recipient of the prestigious Premchand Roychand Scholarship from Calcutta University. In 1943, S. K. Chakrabarty obtained the D.Sc. degree from Calcutta University for his original research contributions in Mathematical Physics. He was awarded the 'Mouat Medal' by the University of Calcutta for his pioneering work in the field of 'Cascade Theory'. In 1944 the Royal Asiatic Society of Calcutta awarded him the 'Elliot Prize'. In 1945 he left Calcutta and moved over to Bombay. Subsequently he worked in USA at the California Institute of Technology, Pasadena as a Visiting Research Fellow. There he started doing research on 'Theoretical Seismology' with eminent seismologists such as Prof. B. Gutenberg, Dr. Benioff and Prof. C. F. Richter. After returning to India, he served as a Professor and Head of the Department of Mathematics in Bengal Engineering College, Shibpur, West Bengal for fourteen long years from 1949 to 1963. In 1963 he became the 'Rashbehary Ghosh Professor' and Head of the Department of his alma mater, the Applied Mathematics Department of Calcutta University. He served in that capacity till his retirement in 1974. In India he in collaboration with H. J. Bhabha was a pioneer research worker in the area of 'Cascade Theory'. He initiated research on 'Theoretical Seismology' in India. He was a giant star in the galaxy of applied mathematicians of Bengal. He published about 50 original research papers in well-known national and international journals.

13. Prof. Manindra Chandra Chaki (M. C. Chaki) was a student of the Department of Pure Mathematics. In 1936 he passed the M.Sc. examination in Pure Mathematics with a First class. After working in a number of undergraduate colleges in present-day Bangladesh [erstwhile East Pakistan] for many years, in 1952 M. C. Chaki was finally appointed as a Lecturer in his alma mater, the Department of Pure Mathematics, Calcutta University. In 1972 he was selected as the 'Sir Asutosh Birth Centenary Professor of Higher Mathematics' (formerly known as 'Hardinge Professor of Pure Mathematics') in Calcutta University and adorned that post till his retirement in 1978. Professor Chaki has made notable contributions in the fields of Riemannian geometry, Classical and Modern Differential geometry, Theoretical Physics, General Relativity, Cosmology and History of Mathematics. A notable contribution of Prof. Chaki relates to his introduction of the notion of Pseudo-Symmetric Manifolds in 1987. In modern geometrical literature this is now known as 'Chaki Manifold'. He published more than 60 original research papers in noted national and international journals. He guided more than 20 research students for their Ph.D. degrees in various branches of Mathematics.

14. Prof. Calyampudi Radhakrishna Rao [C. R. Rao] FRS is an outstanding statistician of India, second only to Prof. P. C. Mahalanobis. He was born to a Telugu family in 1920. He obtained his M.Sc. degree in Mathematics from the Andhra University, Waltair in 1941. He stood First with a First class. Then he joined the newly created Department of Statistics of Calcutta University in 1941 and obtained an M.A. degree in Statistics again standing First with a First class. He obtained record marks in that examination which is still unbeaten. He joined the ISI, Calcutta in 1943 and started working under the supervision of Prof. P. C. Mahalanobis. Later Prof. Mahalanobis sent him to the University of Cambridge to work at the 'Anthropological Museum' there. In Cambridge C. R. Rao started pursuing research under the world famous statistician Prof. R. A. Fisher FRS. In 1948, C. R. Rao obtained his Ph.D. degree from the University of Cambridge for his thesis entitled *'Statistical Problems of Biological Classifications'* under the joint supervision of Professors Fisher and Mahalanobis. On his return to India he joined the ISI, Calcutta and served there for 40 years till he took mandatory retirement at the age of 60. He served ISI in various important capacities including Head and later Director of Research and Training School, Jawaharlal Nehru Professor and National Professor of India. After retirement Prof. C. R. Rao migrated to USA and worked there as Professor and still works in other prestigious posts till date. An elected Fellow of the Royal Society of London (FRS), Prof. C. R. Rao has received innumerable famous and prestigious awards both in India and abroad. He has authored close to 400 research papers and 14 books. For the last seven decades Prof. C. R. Rao has been considered as a leading statistician of the world. His best known scientific discoveries are 'Cramèr-Rao bound' and the 'Rao-Blackwell Theorem'. He has also made notable contributions in the areas of Differential geometry, Multivariate Analysis and Estimation Theory.

15. Prof. Anadi Shankar Gupta [A. S. Gupta] a brilliant product from the Department of Applied Mathematics, Calcutta University and a much admired Faculty Member of IIT, Kharagpur, West Bengal is renowned for his outstanding contributions in the field of Fluid Dynamics, stability of flows on Newtonian and Non-Newtonian Fluids, Boundary Layer Theory and Magneto-Hydrodynamics, notably on heat transfer in free convection flow in the presence of magnetic field. He completed his Master's degree from Calcutta University in 1954. He joined the Department of Mathematics in the Indian Institute of Technology (IIT), Kharagpur as an Assistant Lecturer in 1957. He became a Professor in the same Department in 1968 and continued in that capacity till 1993. In 1972 he received the prestigious Shanti Swarup Bhatnagar Award for his original contributions in Fluid Dynamics and Magneto-Hydrodynamics. He received several awards and was elected Fellows of well-known Indian academies. His personal research contributions are 157 research papers published in reputed national and international journals. He authored a book titled *'Calculus of Variations with Applications'* [Prentice Hall, New Delhi]. He guided a number of Ph.D. students.

16. Prof. (Mrs.) Jyoti Das (nee Chaudhuri) [J. Das] was the greatest lady mathematician not only in Bengal but perhaps in the whole of Eastern India. She was a brilliant product of the Department of Pure Mathematics, Calcutta University. She stood First with a First class in B.Sc. Mathematics (Honours) examination in 1956. She topped the list of all successful candidates in B.A. and B.Sc. Examinations of the University of Calcutta that year. In 1958 she repeated the performance, standing First with a First class in the M.Sc. Examination in Pure Mathematics and simultaneously topping the list of successful candidates at the M.A. and M.Sc. Examinations of the University of Calcutta that year. She obtained her D. Phil degree from the University of Oxford, England for her original research in the Theory of Eigenfunction Expansions. Her areas of specialization comprised Special Functions, Eigenfunction Expansions, Basic and Ordinary Differential Equations. In her long and illustrious career she became the first lady mathematician in India to hold a Chair Professorship in Mathematics. She published 58 original research papers in well-known national and international journals. 8 students obtained their Ph.D. degrees working under her supervision. She authored several books for undergraduate Mathematics courses. But her treatise titled 'Differential Equations' was posthumously published after her demise in 2015.

17. Brief discussions have been made about 7 outstanding mathematical scientists who made remarkable contributions during their stay at ISI, Calcutta. Unfortunately, they left Bengal after some years. They are Professors S. N. Roy, D. Basu, J. K. Ghosh, S. R. S. Varadhan, R. Ranga Rao, K. R. Parthasarathy and V. S. Varadarajan.

- **Discussions on the Famous Mathematical Scientists:**
- **Epilogue**

(i)   Concluding Remarks and Discussions,

(ii) Compiled Bibliographies of the 16 listed mathematical scientists.

Chapter 2
Sir Asutosh Mookerjee (1864–1924)

2.1 Birth, Family, Education

Asutosh Mookerjee was a well-known educationist, a peer-less Vice-Chancellor, an eminent Lawyer, a superlative Judge and also a brilliant mathematician. In this article, I shall primarily concentrate on his contributions as a mathematician.

Asutosh was born in the city of Calcutta on the 29 June 1864 in a well-to-do Western educated Bengali family. His father Dr Gangaprasad Mookerjee was a well-known doctor. One of his uncles Radhika Prasad was an executive engineer. Asutosh's father and uncles were first-generation Western educated professionals. It was a notable transformation from an orthodox Brahmin family of Sanskrit Pundits to a family of doctors and engineers shifted to and settled in Calcutta. Right from his childhood Asutosh was exposed to an intellectual atmosphere. Though he had many tutors, but his real academic training was imparted by his father and two uncles at home.

Asutosh was recognized as a child prodigy in Mathematics even by his school teachers. There are many anecdotes about his passion in handling numbers. To mention just one of these happenings would be interesting. Once as a boy, he was punished by his father for doing some mischief and was locked up in a room. Several hours later, when the door of the room was unbolted, it was found that Asutosh had worked out problems of mathematics all over the walls of that room with a mere piece of charcoal.

After a bright school career, in 1879, he stood second in the Entrance Examination (equivalent to Class X final) conducted by the University of Calcutta. He then joined the famous Presidency College in Calcutta and studied there from 1880 to 1885. During the initial years in College, Asutosh was rather unwell. He suffered from chronic headaches and had a serious accident when his right hand was injured in electric shock. However, overcoming all these impediments, he appeared for the F. A. Examination (equivalent to Class XII) and stood third in the University of

© The Author(s), under exclusive license to Springer Nature Singapore Pte Ltd. 2022
P. Mukherji, *Notable Modern Indian Mathematicians and Statisticians*,
https://doi.org/10.1007/978-981-19-6132-8_2

Calcutta. In 1883, he stood First with a First class in B.A. Mathematics from the same University. He received the Ishan and Vizianagram scholarships and the Harishchandra Prize. While in college he used to read a lot of books on various subjects outside his assigned curriculum. He was particularly fascinated by books on Mathematics and Physics. The notable fact is that, he hardly missed any new publication of a book by a famous mathematician. It may be noted that at the age of 17, when Asutosh Mookerjee was only a first-year student in Presidency College, he published a research paper in Mathematics in an international journal. Detailed discussion on his research contributions in Mathematics will be done later.

Clearly, his academic career was one of uniform brilliance. In 1885, he stood First with a First class in M.A. examination in Mathematics from Presidency College (affiliated to the University of Calcutta). In 1886, he passed M.A. in the Natural Sciences, topping the list of successful candidates once again. He was the first student in the University of Calcutta, to obtain the master's degree in more than one subject. In the same year, he was awarded the prestigious Premchand Roychand Scholarship in Mathematics and Science on the basis of result of a competitive examination. Thus he obtained the coveted Blue Ribbon of the University career.

Asutosh Mookerjee was a versatile genius, but his main interest was in the field of Mathematics. In this context, it would be worthwhile to analyse the influences, which inspired him in this direction. Apart from his father and uncles, one of his father's great friend Dr. Mahendra Lal Sircar (1833–1904) was virtually a mentor of young Asutosh. Dr Mahendra Lal Sircar founded the Indian Association for the Cultivation of Science (IACS) in Calcutta in 1876. He and Father Lafont, a well-known Professor of Physics from St. Xaviers College, Calcutta, inspired Asutosh Mookerjee in the pursuit of Mathematics and Physics. In Presidency College, William Booth, John McCann and J. A. Martin were his teachers in Mathematics. These Professors greatly inspired and encouraged young Asutosh in his study and pursuit of Higher Mathematics. These were some of the important influences, which largely contributed to the making of a mathematician in him. These points are being highlighted, because these are the forces, which emboldened Asutosh Mookerjee to venture into the untrodden paths of mathematical research single-handedly. Asutosh could not find a suitable person to guide him in his research. Undaunted by the difficult situation, he still set forth and made very important contributions in the field of Mathematics.

2.2   Research Contributions, Career

During the twelve years from 1880 to 1892, Asutosh was on one hand preparing for and taking his B.A., M.A. and other examinations, and on the other hand he was publishing original and high-quality mathematical papers. As already mentioned earlier, at the age of 17, when Asutosh was only a first-year student at Presidency College, he published his first research paper in Mathematics, entitled '*On a Geometrical Theorem: Proof of Euclid I, 25*' [Messenger of Mathematics, Vol. 10, (1880–1881), pp. 122–123]. Messenger of Mathematics was published from England. In this paper, Asutosh gave an elegant new proof of the 25th proposition of the first book of Euclid. In this context, it is worth mentioning, that while still an undergraduate student in Presidency College, Asutosh was enamoured by the works of various French mathematicians. He was greatly influenced by A. M. Legendre (1752–1833), G. Monge (1746–1818) and others. They inspired him to take up studies in Geometry. Again as an undergraduate student, he published his second paper entitled '*Extension of a Theorem of Salmon*' [Messenger of Mathematics, Vol. 13, (1883–1884)]. In this paper, he gave some extensions of a theorem enunciated by Salmon.

Another French mathematician who was greatly admired by young Asutosh Mookerjee was J. L. Lagrange (1736–1813). Clearly he was fascinated by the mathematical expositions of Lagrange and probably that is what influenced him to take up elliptic functions as an area of research. While still a student of the M.Sc. class, Asutosh wrote his third paper entitled '*A note on Elliptic functions*' [Quarterly Journal of Pure and Applied Mathematics, Vol. 21, (1886), pp. 212–217]. Clearly, the treatises of Lagrange named '*Theorie des Functions Analytiques*' (1797) and '*Lecon Sur Le Calculus des Functions*' (1804) and the works of A. M. Legendre formed the base of Asutosh's research on elliptic functions. This was an outstanding contribution. In this paper he established a certain addition theorem in the theory of elliptic functions using a new method which utilized the properties of the ellipse. The well-known British mathematician A. Cayley (1821–1895) after reading Asutosh's third paper had observed that the paper was remarkable because in it a real result was obtained by consideration of an imaginary point. Alfred Enneper (1830–1885) in '*Elliptische Funktionen*' has referred to this paper. Later on Asutosh Mookerjee again wrote two papers where the elliptic functions were applied to the problems of mean values. The papers entitled '*Some Applications of Elliptic Functions to Problems of Mean Values; First and Second papers*' [Journal of the Asiatic Society of Bengal, vol. 58, pt II, No. 2, (1889) pp. 199–213; pp. 213–231] are two important publications.

Italian mathematician G. Mainardi (1800–1879) was well known for his success in determining the differential equation for oblique trajectory of a system of confocal ellipses. Asutosh Mookerjee became interested in the problem. He worked out interesting derivations and made important interpretations. Mainardi's solution was extremely complicated and cumbersome. It was impossible to trace the curve from it. Asutosh showed by an ingenious process that Mainardi's integral solution could be replaced by a pair of remarkably simple equations. From this, interesting geometrical interpretations could also be made. These elegant equations as established by Sir Asutosh have been incorporated by Professor A. R. Forsyth (1858–1942) in his classical treatise on Differential Equations, in latter editions. Enthused by the success of solving the problem on a particular trajectory so elegantly, Asutosh published another paper on general trajectory. This paper amply displays Asutosh Mookerjee's power of generalization and elegant expression. In his discussions, Asutosh Mookerjee pointed out that, he had to examine whether a proposed interpretation of a given differential equation is relevant or not. He wrote:

Firstly, the interpretation must give a property of the curve whose differential equation we are interpreting; in fact, it must be a geometrical quantity which vanishes at every point of every curve of the system.

Secondly, the geometrical quantity must be adequately represented in the differential equation to be interpreted.

The research papers that Asutosh Mookerjee published in this context are listed below:

1. *'On the differential equation of a trajectory'*. [Journal of the Asiatic Society of Bengal, (1887), 56, pp. 117–120]
2. *'A memoir on plane analytical geometry'*. [Journal of the Asiatic Society of Bengal, (1887), 56, pp. 288–349]
3. *'A general theorem on the differential equations of trajectories'*. [Journal of the Asiatic Society of Bengal, (1888), 57, pp. 72–99]

Influenced by another French mathematician G. Monge (1746–1818), Asutosh began doing research by coupling Geometry and Calculus, and this resulted in several very important publications. He made some advancement towards Monge's rule related to general differential equations of second degree in x and y which represents a conic. Monge came up with a fifth-order nine-degree differential equation. Though the equation became general the corresponding geometrical interpretation became very complicated. It is noteworthy, that without any guidance from anywhere, depending exclusively on his own brilliant mind, Asutosh successfully gave geometrical interpretation of Monge's Differential Equation to all conics. His interpretation was that '*the radius of curvature of the aberrancy curve vanishes at every point of every conic.*' Edwards in his book 'Differential Calculus' has quoted this interpretation.

It may be noted that Asutosh Mookerjee's paper roused a lot of interest among a section of mathematicians. In a letter written to Asutosh Mookerjee from Cambridge, dated 14 September 1887, A. Cayley in a way supported Asutosh's criticism of J. J. Sylvester's interpretation of the Mongian and wrote: '*it is of course all perfectly right*'. Col Cunningham wrote '*Professor Asutosh Mukhopadhyay has proposed a really excellent mode of geometric interpretation of differential equation in general ....*'

Asutosh Mookerjee kept himself very well informed about the British School of Mathematics, and he acknowledged the professional support he received from the British mathematicians. A. Cayley (1821–1895) and J. J. Sylvester (1814–1897) were very helpful to Asutosh Mookerjee in the professional field.

But the research conducted by the French School of Mathematics especially those with geometrical leanings were very close to his heart. S. D. Poisson (1781–1840) was a notable mathematician of France who made very useful contributions in the field of Theoretical Mechanics. His methods of integration gave alternative forms of differential equations. From Sir Asutosh's 'Diary' we find that he had read two of Poisson's books '*Theorie Mathematique de la Chateur*' (1835) and '*Nouvelle Theorie de L'action Capillarie*' (1831) very extensively. Clearly that inspired him to look closely into Poisson's work and specially his stand regarding analytic and synthetic approaches to Geometry. While evaluating the well-known Poisson's integral which was first considered by S. D. Poisson in his memoir '*Suite du Memoire Sur Les Integrales Definies*' using a technique different from what Poisson did, Asutosh Mookerjee actually obtained a formula of transformation. This method also had the advantage of showing how the indefinite integral itself may be evaluated. In this process of evaluation he arrived at a symbolic value of Π. With his well-known love for Geometry, in this paper also, Asutosh Mookerjee tried to give geometric interpretation to Calculus. It may be mentioned in this connection that the paper entitled '*On Poisson's Integral*' [Journal of the Asiatic Society of Bengal, (1888), 57, 100–106] is a remarkable contribution.

The general impression about Sir Asutosh, the mathematician, is that he was obsessed with Calculus and Geometry. But the myth is not acceptable because of the following two publications by Sir Asutosh Mookerjee. They are

(1) '*On Clebsch's Transformation of the Hydrokinetic Equations*' [Journal of the Asiatic Society of Bengal, (1890), 59, pp. 56–59]
(2) '*Note on Stoke's Theorem and Hydrokinetic Circulation*' [Journal of the Asiatic Society of Bengal, (1890), 59, pp. 59–61]

These are research papers on Fluid Mechanics, a very important branch of Applied Mathematics. Asutosh Mookerjee's interest in physical phenomena led him to the serious study of the works of the German mathematician R. F. A. Clebsch (1833–1872). Clebsch made important contributions in the general theory of curves and surfaces, their uses in Geometry, in the theory of invariants and in elliptic functions. The theories that he developed were applied by him to various physical problems. Asutosh was greatly impressed by such exercise and was drawn to Hydrokinetics.

In his student days in Presidency College, Asutosh Mookerjee had already read the classical treatise of Fluid Dynamics by H Lamb, namely '*A Treatise on the Mathematical Theory of Motion of Fluids*'. So in the paper entitled '*On Clebsch's transformation of the hydrokinetic equations*', [Journal of the Asiatic Society of Bengal, Vol. 59, pt. II, No: 1, (1890), pp. 56–59], he considered hydrokinetic equations in three cases:

i.   Irrotational motion
ii.   Steady rotational motion
iii. General rotational motion.

He demonstrated how the method of Clebsch's transformation to the third case may be simplified. In the second paper of this series, entitled '*Note on Stokes's Theorem and Hydrokinetic Circulation*' [Journal of the Asiatic Society of Bengal, (1890), 59, pp. 59–61], Asutosh Mookerjee gave a new proof of Stokes's formula of hydrokinetic circulation using Clebsch's transformation. Incidentally, he himself has remarked in the second of these papers, '*It is worth noting that as no physical conception enters into the above proof, it holds good, whether we regard the theorem as a purely analytic one or as merely furnishing a formula for hydrokinetic circulation*'. Clearly both the papers are more analytical in nature, and his contributions in this regard are as remarkable as those of Clebsch.

Asutosh Mookerjee was a trendsetter and a pioneer mathematical researcher in the field of Fluid Mechanics. Later on B. B. Datta, S. K. Banerjee, N. R. Sen, N. N. Sen and many others followed in his footsteps. Sir Asutosh Mookerjee remained a very powerful influence on the young research students of those times.

He was a trendsetter in the field of research in Geometry as well. Inspired by Sir Asutosh Mookerjee, many young mathematicians of the early twentieth century took up research in various branches of Geometry. Notable among them are Professor Syamadas Mukhopadhyay, Professor R. C. Bose, Professor Haridas Bagchi, Professor R. N. Sen and others.

Asutosh Mookerjee had a unique mind, which could understand the frontline developments in the Mathematical Sciences of those times. His strong grasp of Geometry and clear understanding of Analysis and Calculus gave him a very wide vision of contemporary Mathematics. He wrote a text book on Conics titled '*An Elementary Treatise on the Geometry of Conics*' [Macmillan & Co, London (1893)]. The book was very popular in those days, both among the students as well as teachers of Mathematics. Recently the Cambridge University Press has published the book again. He and his teacher Shyama Charan Basu jointly wrote a book titled '*Arithmetic for Schools*', which was very helpful for school going students.

Even when he was extremely busy in legal or administrative profession, Asutosh kept himself well informed about the latest developments in Mathematics by reading the contemporary books and journals on the subject.

As Vice-Chancellor of Calcutta University, he commissioned Prof W. H. Young, the then Hardinge Professor of Higher Mathematics of the same University to submit a report on the study of Mathematics in the important centres of learning of the European Continent. This demonstrates Sir Asutosh's commitment to keep the Mathematics courses of the University of Calcutta at par with their international counterparts.

2.3 Special Honours

For his original research in Mathematics, Sir Asutosh Mookerjee was honoured by many national and international academic bodies related to the subject. In 1885, at the age of 21, he was elected a Fellow of the Royal Astronomical Society of Edinburgh. In 1886, he was elected as a Fellow of the Edinburgh Royal Society. He also became a Member of the Royal Asiatic Society and Bedford Association for the improvement of Geometrical Teaching.

In 1887, Asutosh was appointed Honorary Professor of Mathematics at the Indian Association for the Cultivation of Science (IACS) for two years. He delivered 30 lectures during those two years and covered topics on Physical Optics, Mathematical Physics and Pure Mathematics.

In Pure Mathematics, Asutosh Mookerjee delivered courses on Analytic Geometry, Boole's Theorem, Boole's Theorem on linear transformations, asymptotes and eccentricity, theorems on central conics, non-central conics, confocal conics and plane elliptic coordinates, theory of analytic functions, integration of algebraic functions, hyperbolic functions, Abel's theorem, Dirichlet's theorem, Gamma functions, etc. The topics mentioned above indicate that Asutosh as a teacher was very familiar with the frontline areas of mathematics, and during his very brief tenure in the IACS, he had impressed the students with his commendable and deep knowledge of the different branches of Pure as well as Applied Mathematics. During that time, he was also appointed examiner in Mathematics by the Calcutta University. In 1887, he was elected Fellow of the London Physical Society.

In 1890, as recognition of his original contribution to Mathematics, he was made a Member of the Mathematical Society of Palermo, Siciliy and Societe de Physique of France.

In 1908, Sir Asutosh founded Calcutta Mathematical Society and became its Founder-President. He nurtured the society very carefully and ultimately the Society became a great and vibrant centre which aided the future researchers of Mathematics a great deal. Asutosh Mookerjee, as a President of the Calcutta Mathematical Society, guided the activities of the Society from its inception till his own death in 1924. During Sir Asutosh's lifetime, 83 meetings of the Society (other than council meetings) were held, and he presided over each one of these meetings. For a busy man like him, it was a rare feat indeed. Such was his dedication to the Society. Under his personal initiative, encouragement and inspiration, research workers in mathematical and physical sciences began to gather around him. Of these scholars later on some became world renowned scientists. These young scholars delivered addresses on their research work before the Members of the Society, and their contributions were published in the pages of the Bulletin of the Calcutta Mathematical Society. Just to give an idea about Asutosh Mookerjee's success in encouraging the talented mathematicians to come forth with their research contributions, it would be sufficient to state that in the first meeting of the Society in 1909, papers were read by Professors C. E. Cullis, Ganesh Prasad, S. D. Mukhopadhyay and P. L. Ganguly. In 1914, B. B. Dutta presented his research work before the Society. In 1915 M. N. Saha and S. K. Banerjee did the same. In 1916, S. N. Bose delivered an address on his research work followed by D. N. Mallik, N. R. Sen and others in 1917. The trend continued under the careful patronage of Asutosh Mookerjee. All the well-known mathematicians of Calcutta which included the above legendary names as well as the younger talents like B. M. Sen, N. N. Sen, J. Ghosh, G. Bhar, B. Sen and others had their inspiration and encouragement from the greatest pioneer in mathematical research in the country, Sir Asutosh Mookerjee.

One is amazed at Sir Asutosh's far sight. In the early twentieth century, he had realized the need for an academic society in Mathematics. He, the true pioneer of mathematical research in the country, left no stone unturned to create a conducive environment of mathematical research for the future generation of scholars. Calcutta Mathematical Society was Sir Mookerjee's gift to the future generations of researchers in mathematical and physical sciences in this country.

Just to demonstrate, how Sir Asutosh Mookerjee was regarded by some of the greatest Indians of those times, the following remarks are very noteworthy. Rabindranath Tagore, the greatest among Indian intellectuals wrote about Asutosh:

'*He had the courage to dream because he had the power to fight and the confidence to win ... his will itself was the path to the goal.*' Regarding his contributions in the field of mathematical research, Dr. Ganesh Prasad wrote '*After Bhaskara he was the first Indian to enter into the field of Mathematical Research as distinguished from Astronomical Research and much which was truly original*'.

After Sir Asutosh Mookerjee suddenly passed away in Patna in 1924 at the age of sixty, Nobel Laureate Sir C. V. Raman lamented and said that Bengal in gaining a distinguished Judge and a great Vice-chancellor lost in him a still greater mathematician.

2.4 Important Milestones in the Life of Sir Asutosh Mookerjee

1864: Born on 29 June, as the eldest son of Gangaprasad Mukhopadhyay and Jagattarini Debi, in Calcutta.

1879: Passed the 'Entrance Examination' with scholarship.

1880–1883: Stood third in First Arts (Intermediate) from Presidency College, Calcutta. Stood First in B.A. with First class in Mathematics from Presidency College, Calcutta.

1886: Passed M.A. in Physical Science and Mixed Mathematics (November 1886). The first student in Calcutta University to obtain a master's degree in more than one subject. He stood First with a First class in both the examinations.

1886: Won Premchand Roychand Studentship and Mouat Medal. Elected Fellow, Edinburgh Royal Society; Member, Royal Asiatic Society and Bedford Association for the improvement of Geometry Teaching.

1887: Appointed Honorary Professor of Mathematics at the Indian Association for the Cultivation of Science (1887–1890). Appointed examiner in M.A. in Mathematics, Calcutta University. Became Fellow, London Physical Society.

1888: Elected Fellow, Mathematical Societies of Edinburgh and Paris.

1890: Member, Mathematical Society of Palermo, Sicily and Societe De Physique of France.

1893: Elected as Member of the Irish Academy.

1890: Elected Fellow, American Mathematical Society.

1906: Appointed Vice-Chancellor of the University of Calcutta and held office for four successive terms (1906–1914).

1908: Founder—President of the Calcutta Mathematical Society. Conferred D.Sc. (*Honoris Causa*) by the University of Calcutta. President, Indian Association for the Cultivation of Science (IACS), Calcutta.

1909: Appointed Companion of the Order of the Star of India (CSI) by the ruling British Government.

1911: In December 1911 he was conferred 'Knighthood' by the ruling British Government.

1913: Established Post-Graduate Departments with teaching and research facilities at Calcutta University.

1914: Foundation stone of the College of Science of Calcutta University was laid by him on the 27 March 1914.

1914: Founder-President of the Indian Science Congress Association.

1921: Appointed for the fifth time the Vice-Chancellor of the University of Calcutta (1921–1923).

1924: In May 1924 died suddenly at Patna, Bihar.

1964: The Government of India issued postage stamp on the occasion of his 'Birth Centenary' to honour him for his contributions as an educationist per excellence.

Chapter 3
Syamadas Mukhopadhyay (1866–1937)

3.1 Childhood, Education

Syamadas Mukhopadhyay (S. D. Mukhopadhyay) was born at Haripal in the Hooghly district of West Bengal on 22 June 1866. His father Babu Ganga Kanta Mukhopadhyay being in the state judicial service had different places of posting, and as a consequence Syamadas had to study in different institutions at different times. On completion of his graduation from Hooghly College he obtained his M.A. degree in Mathematics from the Presidency College, Calcutta, in 1890. For his mathematical dissertation entitled '*On the infinitesimal analysis of an arc*', he was awarded the Griffith's Prize of Calcutta University in 1909. In 1910, for his pioneering work on Differential Geometry, he was awarded the Ph. D. degree by the University of Calcutta, and the name of his thesis was 'Parametric Coefficients in Differential Geometry of Curves'. S. D. Mukhopadhyay was the first Indian recipient of the Ph.D. degree in Mathematics from Calcutta University.

3.2 Career, Research Contributions

Prior to his appointment in government service as a Professor of Mathematics in the Bethune College, Calcutta, Syamadas Mukhopadhyay joined a private college in Calcutta as a Faculty Member soon after completing his M.A. in Mathematics and served there for several years. In Bethune College, the first women's college in Asia, S. D. Mukhopadhyay had a very heavy teaching load, and apart from Mathematics, he had to take classes there on English and Philosophy regularly. In 1904, he was transferred to the Presidency College. He served there for eight years till 1912. In 1912, Syamadas Mukhopadhyay was invited to join the newly set up Department of Pure Mathematics in the Calcutta University by Sir Ashutosh Mookerjee, the then Vice-Chancellor of Calcutta University. S. D. Mukhopadhyay accepted the offer and joined the department. He served there to the satisfaction of Sir Ashutosh and

P. Mukherji, *Notable Modern Indian Mathematicians and Statisticians*,
https://doi.org/10.1007/978-981-19-6132-8_3

continued his service till 1932, and during the last six years of his tenure, Syamadas held the post of a University Professor.

As a talented and eminent teacher of Pure Mathematics, name of S. D. Mukhopadhyay spread everywhere. In order to be a direct student of Professor Mukhopadhyay, R. C. Bose after completing his M.Sc. in Applied Mathematics from the Delhi University came and joined the Pure Mathematics Department of Calcutta University as a student. He later became a researcher under S. D. Mukhopadhyay and earned eminence as a great mathematician. Among many students of Syamadas Mukhopadhyay, three deserve special mention. They were Professor G. Bhar, Professor R. N. Sen and Professor M. C. Chaki. Professor Bhar was a teacher of great repute in Presidency College, Professor Sen became the Hardinge Professor of Pure Mathematics in Calcutta University and also a famous researcher in the field of Differential Geometry, and Professor Chaki held the Chair named 'Sir Ashutosh Birth Centenary Professor of Higher Mathematics' in Calcutta University and also had a number of research publications in many areas of Geometry.

William Booth, a teacher of Hooghly College, who was quite well known for his researches in Geometry possibly, inspired Syamadas Mukhopadhyay during his undergraduate student days. Professor Booth was the man who had influenced young Ashutosh Mookerjee during the latter's student days in the Presidency College. M. C. Chaki, a famous mathematician of Calcutta University, in his article on Syamadas Mukhopadhyay, wrote that 'In a very short time after taking his B.A. degree he (S. D. Mukhopadhyay) solved an intricate geometrical problem, a fact recorded in M'Celland's 'Geometry of the Circle'. He also mentioned that S. D. Mukhopadhyay's research contributions 'were outstanding for their originality and novelty of treatment'.

Syamadas Mukhopadhyay's research work may be broadly classified into 4 parts; use of Synthetic Geometry to solve properties of plane curves; Non-Euclidean Geometry; Differential Geometry and Stereoscopic representations of four-dimensional space.

In the first part, he has dealt with properties of plane curves, especially in their infinitesimal regions using synthetic methods. Here he had developed new method which led to a number of interesting theorems on the existence of minimum number of cyclic and sextactic points between two points of a given curve on a convex oval, etc. In this context two of his theorems deserve special attention. But before that it would be necessary to understand what are cyclic and sextactic points. A cyclic point is a singular point on a plane curve, where the circle of curvature passes through 4 consecutive points instead of 3. On the other hand a sextactic point is a singular point, where the osculating conic passes through 6 consecutive points instead of 5. At a cyclic point, the circle of curvature may touch the given circle internally or externally. In the former case, the point will be called in-cyclic and in the latter case ex-cyclic. Similarly at a sextactic point, the osculating conic may touch the given curve internally or externally. In the former case, the point is called in-sextactic, and in the latter case it is called ex-sextactic. With these definitions, S. D. Mukhopadhyay first demonstrated a number of interesting propositions. Later they led to his famous two theorems stated below:

Theorem I states that 'the minimum number of cyclic points on a convex oval is 4', and Theorem II states that 'the minimum number of sextactic points on a convex oval is 6'. These two theorems were first published in 1909 in the Bulletin of the Calcutta Mathematical Society (BCMS). But initially, at that time not much attention was paid to this piece of research. Only the eminent French mathematician J. S. Hadamard (1865–1963) referred to this work in the memoirs of College de France. However, much later, these theorems were re-discovered in Europe. Since then many noted mathematicians have taken them up as subject of investigations. W. Blaschke, a noted German geometer, gave credit to Syamadas Mukhopadhyay for giving the original first proof of the Theorem I stated above. In modern literature of Geometry, this theorem is now eponymously quoted as 'Mukhopadhyay's Four Vertex Theorem'. This celebrated theorem is stated in various important mathematical treatises such as *'Differential Geometry'* by H. Guggenheimer [McGraw Hill, New York, 1930]; *'Differential Geometry'* by J. J. Stoker [Wiley Inter Science, 1969] and *'Elements of Differential Geometry'* by R. S. Millman and G. D. Parker [Prentice Hall, Englewoods Cliffs, 1977].

As already mentioned, W. Blaschke gave the credit of proving the 'Four Vertex Theorem' to Syamadas Mukhopadhyay. He has mentioned this in his book *'Vorlesungen Uber Differential Geometry'* [Springer, Berlin, 1924]. Further references connected to this theorem are also given there. Later on S. D. Mukhopadhyay generalized these two theorems. The generalized Theorem I states that 'If a circle C intersects an oval V in $2n$ points ($n < 2$) then there exists at least $2n$ cyclic points in order on V, of alternately contrary signs, provided the oval has continuity of order 3'. The generalized Theorem II states that 'If a conic C intersects an oval V in $2n$ points ($n >$ or $= 2$), then there exist at least $2n$ sextactic points in order on V, which are alternatively positive and negative, provided V has continuity of order 5'. Thus S. D. Mukhopadhyay placed his earlier investigations on more rigorous basis.

Apart from these investigations, Syamadas Mukhopadhyay published a number of research papers on the general theory of osculating conics. In this context, the comments made by his student G. Bhar of Presidency College may be noted. Bhar wrote '*It is highly probable that Mukhopadhyay was led to the study of the theory of osculating conics by Sir Ashutosh's work on the differential equations of all parabolas and his beautiful geometrical interpretation of the Mongean equation*'.

In the second part of his research work, Syamadas Mukhopadhyay contributed to a type on Non-Euclidean Geometry namely Hyperbolic Geometry. Actually Hyperbolic Geometry is the geometry obtained by assuming all the postulates of Euclid except the fifth one. The fifth postulate of Euclid is replaced by its negation. In the context of S. D. Mukhopadhyay's research in Non-Euclidean Geometry, the comments made by his student G. Bhar are very relevant which states:

Professor Mukhopadhyay found himself in his elements when he was called upon to teach the principles of Non-Euclidean Geometry to the Post Graduate students of the University.

Serious research work in Non-Euclidean Geometry in India was first undertaken by R. Vaithyanathaswamy (1894–1960) of Madras University. His first two research papers in this field were published in the *Journal of the Indian Mathematical Society*

(JIMS) in 1914. The first two research papers of Professor Vaithyanathaswamy were titled *'Parallel straight lines'* [Journal of the Indian Mathematical Society, 6(1), (1914), 58–61] *and 'Length of a circular arch'* [Journal of the Indian Mathematical Society, 6(1), (1914), 220–221]. These were in all likelihood the first research papers on Non-Euclidean Geometry published by an Indian.

S. D. Mukhopadhyay in collaboration with his student G. Bhar published his first research paper on Hyperbolic Geometry in the Bulletin of the Calcutta Mathematical Society in 1920. The paper was titled *'Generalisation of certain theorems in the Hyperbolic Geometry of the triangle'*. This publication was followed by a number of research papers on this kind of Geometry. The investigations carried out by him resulted in the important discovery of the 'Rectangular Pentagon' and his beautiful geometrical interpretation of the Engel-Napier rules. In collaboration with his two famous students R. C. Bose and G. Bhar he made interesting generalizations of the ideas of concurrence and collinearity of lines and points in Non-Euclidean Geometry. To be more specific, S. D. Mukhopadhyay extended the well-known concurrency theorems of the angle bisectors and the right bisectors of the sides of an ordinary triangle, to all types of hyperbolic triads of lines and points. It may be noted that by a hyperbolic triad, it is meant a group of three elements (points or lines) lying upon a hyperbolic plane. In this context, his three research papers titled *'Geometrical investigations on the correspondence between a right angled triangle, a three-right-angled quadrilateral and a rectangular pentagon in hyperbolic geometry'* [Bulletin of the Calcutta Mathematical Society, 13(4), (1922–1923): 211–216]; *'On general theorems of co-intimacy of symmetries and hyperbolic triads'* [Bulletin of the Calcutta Mathematical Society, 17(1), (1926: 39–55] and (with R. C. Bose) *'Triadic equations in hyperbolic geometry'* [Bulletin of the Calcutta Mathematical Society, 18, (1927), 99–110] deserve special attention.

In the first paper mentioned above, S. D. Mukhopadhyay concluded 'We have thus the closed series of 5 associated right-angled triangles and the Engle-Napier Rules are shown to possess a real geometrical basis in the rectangular pentagon'. It was an exquisite piece of mathematical research.

In the third part of his research and investigations, Professor Syamadas Mukhopadhyay while dealing with Differential Geometry of curves introduced for n-dimensional space curves certain differential forms. He named them parametric coefficients, and using them he expressed many invariant properties of the curves. In this matter, eminent geometer M. C. Chaki has commented 'The distinct merit of the method of parametric coefficients of Mukhopadhyay lies in the fact that it achieves by elementary method, results which have been obtained by advanced analysis.'

The fourth and final part of his research comprises a suggestion of a stereoscopic device for visualizing figures in four-dimensional space and his discussion with Bryan in this matter.

In 1912–1913, S. D. Mukhopadhyay had published his note on the stereoscopic representation of four-dimensional space [Bulletin of the Calcutta Mathematical Society, 4 (1912–1913), 15]. Around the same time Bryan had also published a paper in the same journal suggesting another kind of device. He had also claimed that it was superior to the one suggested by S. D. Mukhopadhyay. A reply to this criticism was also given by Syamadas Mukhopadhyay, and it too was published in the Bulletin of the Calcutta Mathematical Society [Bulletin of the Calcutta Mathematical Society, 6, (1914), 55–56]. In that reply Syamadas Mukhopadhyay wrote: 'I do not however see any good in further prolonging the controversy between us. Both of us have fairly stated our methods. It would lie with other mathematicians interested in this problem of four dimensions to accept or reject either'.

Towards the end of his research career, S. D. Mukhopadhyay published two important papers entitled 'Lower segments of M-curves' and 'Cyclic curves of an ellipsoid' in the Journal of Indian Mathematical Society. Mukhopadhyay explained that the name M-curve (Monotropic Curve) was due to the German geometrician Stackel H. Mohrmann (1907–1934), who in adopting the name gave the following precise definition. He said 'A singularity-free (that which does not intersect itself) real branch of an analytic curve, which divides the Euclidean plane into two and only two regions, and for which the curvature at every finite point is limited and different from zero, will be called a limited monotropic curve or simply an 'M-curve' [Mathematische Annalen]'.

In his investigations, S. D. Mukhopadhyay considered M-curves which are not necessarily analytic. He has characterized M-curves in a different way and investigated the associated problem.

3.3 Impact of Research (Reviews and Accolades)

W. Blaschke with regard to the paper *'Cyclic curves of an ellipsoid'* suggested in his lecture on *'Selected Problems of Differential Geometry'* in 1932 and said that the object of Dr. Mukhopadhyay's paper is to study certain properties of cyclic curves on an ellipsoid, which is known to possess six vertices.

J. S. Hadmard of the Institute of France had a fair amount of interest in the researches carried out by S. D. Mukhopadhyay. In a letter dated 23 February 1923, he opined on Professor Syamadas Mukhopadhyay's contributions in Synthetic, Non-Euclidean and Differential Geometry and wrote *'The interest of your researches on osculating conics, on even Non-Euclidean Geometry and parametric formulae in differential geometry of curves have been increased by comparison with memoirs published in a slightly different line by a Dane, Juel. Indeed the conjunction of both kinds of works (Juel dealing with straight lines, you with circles and conics), is likely at my seminar, to prove of great power and bearing for further improvement of geometry.'*

Professor Syamadas Mukhopadhyay published 30 original papers in different national and international journal. Barring a few, most of them were published by University of Calcutta as *'Collected geometrical Papers by Professor Syamadas Mukhopadhyay'* in two parts. The volumes were simultaneously reviewed in high-end journals like Nature and the Bulletin of the American Mathematical Society and received favourable comments. The original texts from the journals are quoted below.

The review of the Part I of the collection, as published in Nature of 1931, stated *'The papers in this collection number ten on plane curves and seven on non-Euclidean, mainly hyperbolic geometry. The papers of the first group include six dealing such topics as the geometrical theory of a plane non-cyclic arc, cyclic and sextactic points and a generalized form of Bohmer's theorem, in which methods of pure geometry are employed, in several cases new methods of considerable interest. In this group there are also four papers on the general theory of osculating conics, in which the methods of differential geometry are applied in rather a novel manner. The papers in the second group also offer some new features, and amongst a number of interesting results may be noted an extension of the well known correspondence between a right-angled triangle and a three right-angled quadrilateral in hyperbolic geometry, so as to include a regular pentagon. The book can be recommended to all who are interested in geometry, whether Euclidean or not, and wish to learn something of the progress of the geometrical studies in Indian Universities.'* [Nature, no. 3205, vol. 127, April 4, (1931), 516]

The same collection of papers was reviewed by Professor Virgil Snyder, in the *Bulletin of the American Mathematical Society* in 1931, and the exact text of the review is quoted below. Professor Synder wrote

> *The present collection contains 17 papers, previously published in Asiatic periodicals in the interval 1908–1928. The topics considered fall into three general heads, those concerning topological questions, including cyclic points, sextactic points, etc. of plane curves, those concerning triangles and quadrilaterals in hyperbolic geometry, and those on methods of visualizing representations of four way space. The methods are mostly those of elementary geometry, although differential expressions are used freely in the treatment of osculation. The proofs are strikingly direct and simple, and many of the theorems were first published previously to those obtained by others. For workers in topology, the Papers will be of real service.* [Bulletin of the American Mathematical Society, 36(no.9), (1931), 614]

The second volume of the collected papers was again reviewed in *Nature* in 1932. The review is quoted exactly as the published version was. It said

> *This volume is a continuation of the volume of papers by the same author published in 1929 and reviewed in Nature of April 4, 1931, p. 516. There are two papers on plane convex ovals, but the chief part of the book consists of seven papers on the differential geometry of curves in an N-space. The latter are of special interest both on account of the original methods employed and the results obtained. They deal with parametric coefficients and their properties, the extension of the Serret–Frenet formulae to curves in N-space, the expression of the co-ordinates in terms of the arc, curvatures at a singular point and osculating spherics. Unfortunately, the investigation is restricted to Euclidean space, but the author claims that, by the use of a certain distance formula, it can be adapted to any kind of non-Euclidean space without insurmountable difficulties. It would be of some interest if such a programme were actually carried out, if only for the four- and five-dimensional spaces used in relativity theory. The abstract nature of the topics dealt with makes the papers difficult to read, but students of algebraic geometry should find much to interest them. The book is clearly printed and unusually free from misprints, and is a credit to the Calcutta University Press.* [Nature, no. 3323, vol. 132, July 7, (1933):48]

The review of the same collection of papers was done by Professor Virgil Snyder in the *Bulletin of the American Mathematical Society* in 1932. He wrote *'Part I of the collected papers was published in 1930, and reviewed in this Bulletin, vol. 36 (1931), p. 614. In Part II, the pagination continues, and the make-up is the same as that of Par I. It contains two essays on plane topology which appeared in the Mathematische Zeitschrift in 1931 and the Tohuku Mathematical Journal in 1931 respectively, and seven on parametric representation of curves in n-space, all but one of which (the Griffiths Memorial Prize essay of 1910) were published in the Bulletin of the Calcutta Mathematical Society from 1909 to 1915. The argument and point of view of the papers on topology are similar to those in Part I. Only elementary methods are employed, but with striking originality and richness in new results. Most of these concern cyclic and sextactic points on continuous ovals.*

The other essays are on differential geometry on analytic curves in a Euclidean n-space. Properties are expressed in terms of determinants of derivatives of various orders of the coordinates as to the parameter.

The first intrinsic parameter is the arc length. The second is the projection of the area of the triangle formed by three points which approach coincidence on the curve, summed over the interval of integration, etc. A curve in S has n such intrinsic parameters. They are independent of the coordinates chosen and of the parameter. Any $n-1$ independent equations connecting these parameters will determine a curve in S, intrinsically. The generalised idea of curvature, spheric of osculation, quadric of osculation etc. can now be expressed. The results in the case of plane curves are compared with those obtained by projective differential geometry. The same ideas are then extended to curves in S. At times the amount of machinery necessary seems a bit bewildering, but one is soon consoled by an unexpected general theorem evolving from the maze of formulas. The various kinds of singular points and the associated parametric representation in series are treated in great detail. The papers contain a powerful weapon with which to attack metric problems in analytic curves of hyperspace' [Bulletin of the American Mathematical Society, 38(7), (1932), p. 480].

Some opinions expressed by internationally famous mathematicians about Professor Mukhopadhyay's work may be of interest.

Professor J. Hadamard (Paris) said:

My interest in your new methods in the geometry of a plane arc, which I had expressed in 1909 in an (anonymous) note in recue generale des sciences, has far from diminished since that time.

Precisely at my seminaire or colloquium of the college de France, we have reviewed such subjects and all my auditors and colleagues have been keenly interested in your way of researches which we all consider as one of the most important roads opened to Mathematical Science.

Professor F. Engel (Geissen) commented:

I am surprised over the beautiful new calculations on the right angled triangles and the three-right-angled quadrilaterals (in hyperbolic geometry).... Your analogies in the Gaussian Pentagramma Mirificum are highly remarkable.

Professor W. Blaschke (Hamburg) wrote:

I am much obliged to you for your kind sending of your beautiful geometrical work. When, as I hope, a new edition of my Lessons in Differential Geometry comes out, I shall not forget to mention that you were the first to give the beautiful theorems on the number of Cyclic and Sextactic points on an oval.

Professor F. Cajori (California) wrote to Professor S. D. Mukhopadhyay:

I congratulate you upon your success in research. If ever I have the time and opportunity to revise my History of Mathematics I shall have occasion to refer to your interesting work.

Professor T. Hayashi (Japan) wrote:

Sincerely I congratulate your New Methods in Geometry, *specially on the* concept of intimacy.

Professor A. R. Forsyth (London) wrote:

Your papers connected with analytical and differential geometry are valuable and interesting.

Professor L. Gedeaux (Liege) commented:

A first reading of your papers has seized my grand interest. As I have written, I intend making an exposition of these questions early to my students of Géométrie supéricure, an exposition in which I reckon to join that of the works of M. Juel.

The reviews from internationally reputed journals quoted above, as well as the comments made by the noted mathematicians of the time, clearly reflect the importance of the research conducted by Syamadas Mukhopadhyay and the interest that it generated in the mathematical arena of the thirties of the twentieth century. After his retirement from the Department of Pure Mathematics in 1932, S. D. Mukhopadhyay availed of the Ghosh Travelling Fellowship of the University of Calcutta and proceeded to Europe to study the methods of education followed there.

This also gave him a chance to personally interact with the eminent mathematicians of Europe. He utilized this opportunity and expounded his invented principles to the mathematical circles of the West. He also studied carefully the methods of education prevalent in the countries of Europe he visited. On his return to India he wrote extensive memoirs based on his observations. After the death of Ganesh Prasad in 1935, Syamadas Mukhopadhyay was elected the President of the Calcutta Mathematical Society. He took a keen interest in all matters connected with the Society till his death in 1937.

Apart from the mathematician Syamadas Mukhopadhyay, a closer view of the man himself is very nicely portrayed by his eminent student G. Bhar. He wrote

Mukhopadhyay was a man of broad sympathies and wide culture. He had an artistic bend of mind and was one time a keen amateur photographer of no mean order. His love for the beautiful and the sublime found expression in his ardent passion for rose culture. His collection of roses in his country house in Mihijam is unique in India, and he enriched it every year by directly importing rose trees from England, France, Holland and other European countries.

He was man of simple habits and a very private person who preferred to live away from the glamour of public life. He remained absorbed in his own work. Very correctly his other famous student M. C. Chaki observed '*Mukhopadhyay left an example of plain living and high thinking*'. On 8 May 1937 he breathed his last.

This article is written in an effort to highlight and bring to the knowledge of the present generation the achievements of the great geometrician of India and also to preserve for posterity the memory of a man who contributed so handsomely to the world of Mathematics.

3.4 Important Milestones in the Life of Professor Syamadas Mukhopadhyay

1866: Born on 22 June 1866 at Haripal in the Hooghly district of undivided Bengal.

1888: Graduated from Hooghly College.

1890: Obtained his M.A. degree in Mathematics from the Presidency College, Calcutta (affiliated to the University of Calcutta).

1904: Transferred to the famous Presidency College, Calcutta. He served there for eight years.

1909: Published the original research paper titled '*New Methods in the Geometry of a Plane Arc I, Cyclic and Sextactic Points*' [Bulletin of the Calcutta Mathematical Society, Vol. 1, (1909), 31-37]. This led to his developing the famous theorem, internationally known as '*Mukhopadhyay's Four Vertex Theorem*'.

1909: He won the Griffith's Prize of Calcutta University for his mathematical dissertation entitled '*On the infinitesimal analysis of an arc*'.

1910: He was awarded the Ph.D. degree by the University of Calcutta for his original work in Differential Geometry. His thesis was titled '*Parametric Coefficients in Differential Geometry of Curves*'. He was the first Indian to obtain a doctorate degree in Mathematics in Calcutta University as well as in India.

1912: Joined the newly set up Department of Pure Mathematics of the University of Calcutta as Lecturer.

1926: Became a full Professor in the Department of Pure Mathematics, Calcutta University.

1931–1932: University of Calcutta published most of his important research papers in two volumes titled '*Collected Geometrical Papers by Professor Syamadas Mukhopadhyay*'. The collection received wide admiration from internationally famous mathematicians, and very good reviews were published in '*Nature*' and in '*Bulletin of the American Mathematical Society*'.

1935: Elected the President of the Calcutta Mathematical Society.

1937: He passed away in Calcutta on 8 May 1937.

Chapter 4
Ganesh Prasad (1876–1935)

4.1 Birth, Early Education

Ganesh Prasad was born on the 15 November 1876, at the small town of Ballia in
U.P. He belonged to a well-to-do family. After passing his matriculation examination
in 1891 with a first division, he studied at the Muir Central College, Allahabad. From
there he graduated with a First class Honours in Mathematics in 1895. Subsequently,
he took M.A. degrees in Mathematics from the Universities of Allahabad and Calcutta
and in 1898 was awarded the D.Sc. degree from the Allahabad University for his
original contributions in mathematical research. After a brief stint as lecturer at the
Kayasth Pathshala, Allahabad, in 1899 he obtained a Government of India scholarship
to go to England for pursuing higher studies and research in Mathematics.

4.2 Higher Education, Career, Research Contributions

At Cambridge, Ganesh Prasad studied Mathematics at the Christ Church College with
eminent mathematicians such as Professor E. W. Hobson FRS (1856–1933), A. R.
Forsyth FRS (1858–1942) and Sir E. T. Whittaker (1873–1956). While at Cambridge
he had Allan Baker and Lamour as his contemporaries. At that time he sat for the
Adam's Prize competition of Cambridge University. For this he had submitted his
original research paper entitled '*On the constitution of matter and Analytical Theories
of Heat*'. But that year no one was awarded the prestigious Prize.

Thereafter Ganesh Prasad moved on to Göttingen in Germany to continue his
mathematical research. He started working under the guidance of the famous German
mathematician Professor Felix Klein (1849–1925). He showed his earlier research
paper to Professor Klein. Professor Klein was much impressed by the work and got
the paper published in the Göttingen Abhandlungen ['*Constitution of Matters and
Analytical Theories of Heat*': Abhandlungen d.k Geselschaft der Wiss Zu Gottingen,

© The Author(s), under exclusive license to Springer Nature Singapore Pte Ltd. 2022
P. Mukherji, *Notable Modern Indian Mathematicians and Statisticians*,
https://doi.org/10.1007/978-981-19-6132-8_4

1903] While carrying out research in Göttingen University, Ganesh Prasad came into close contact with men like A. J. Sommerfeld, David Hilbert and Georg Cantor.

During his stay in Göttingen, Ganesh Prasad solved another important problem and wrote it down as a mathematical research paper entitled '*On the Notion of Lines of Curvature*' and showed it to Professor D. Hilbert. Professor Hilbert was quite impressed with Dr. Prasad's method of solution and immediately communicated it to the Royal Society of Sciences of Göttingen. This paper was published in the 'Proceeding of the Royal Society of Göttingen' in 1904.

After five very fruitful years of study and research abroad, Dr. Ganesh Prasad returned to India in 1904. On his return, he was appointed additional Professor of Mathematics at the Muir Central College, Allahabad. Shortly after that he was appointed Professor of Mathematics at the Queen's College, Banaras the same year. In 1905, Pandit Sudhakar Dwivedi, the then Head of the Department of Mathematics at the Queen's College, Banaras retired. Then Professor Ganesh Prasad had to take up additional responsibilities there. Being the only Faculty Member in the department, he had to teach in all the four degree classes single-handedly. But even at that young age and with so much of teaching load, Professor Ganesh Prasad found time to help two students in preparing for their D.Sc. degree. This clearly indicates Professor Ganesh Prasad's great love for mathematical research. This point will be discussed in greater detail and highlighted later.

In spite of his various commitments and a super-hectic work schedule, Sir Asutosh Mookerjee kept himself well informed about contemporary mathematical research of his time. Obviously Ganesh Prasad's achievements in mathematical research and his publications in scientific journals had caught the attention of Sir Asutosh, who was then the Vice-Chancellor of Calcutta University. He was just setting up the Post-Graduate departments of Pure and Mixed Mathematics in the University of Calcutta. So Sir Asutosh lost no time and in 1914, Ganesh Prasad was invited to become the first Rashbehary Ghosh Professor in the Department of Mixed Mathematics, which later on came to be known as the Department of Applied Mathematics. For four years Professor Prasad served the University of Calcutta but in 1917, he returned to Banaras as a Professor of Mathematics in the Banaras Hindu University and also as Honorary Principal of the Central Hindu College, Banaras. After six years, in 1923, he was once again invited by Sir Asutosh Mookerjee to take up the Hardinge Professorship of Higher Mathematics in the Department of Pure Mathematics. Dr. Ganesh Prasad continued in this post till his death on the 9 March 1935.

This great mathematician enhanced the prestige of Calcutta University in two ways. Firstly his personal research contributions in Mathematics are legendary. He was the author of a fairly large number of research papers, memoirs and reviews. Secondly, he was instrumental in building up two strong research schools in the Departments of Mixed Mathematics and Department of Pure Mathematics respectively.

Dr. Ganesh Prasad's research work can be roughly classified into three main groups. The first group consists of papers on Applied Mathematics, especially in the Theory of Potentials. In this field, Ganesh Prasad used his deep knowledge of the

Theory of Functions of a Real Variable. The researchers prior to him had not considered the cases, in which, under special circumstances, the differential coefficients either became infinite or did not exist. Dr. Ganesh Prasad made a thorough investigation of such cases. A noteworthy contribution to this group was the memoir entitled *'Constitution of Matter and Analytical Theories of Heat'*, which has already been mentioned earlier. This paper is considered as an authoritative solution of a difficult problem in mathematical physics. Felix Klein considered it as a very satisfactory solution.

On the suggestion of Hilbert, he took up studies on curvature of surfaces. He also worked on Potential Theory, where he made many significant contributions. In his paper titled *'On the potential of Ellipsoids'* [Messenger of Mathematics, (1901), 8], he used the method of expansion of series. Actually this is indicative of his later works on summation theorem and asymptotic expansions. The most remarkable feature of this paper is that it furnishes an expansion of any algebraic integral functions in spherical harmonics. Since one is concerned with integral functions, knowledge about the singularities of the integrals is not necessary. Dr. Prasad's method of getting round the improper behaviour of certain parameters is quite remarkable. It applies to the expansion in any function space of any number of dimensions.

Though Professor Ganesh Prasad was by training a pure mathematician, some of his research papers on many problems of applied mathematics are really remarkable. Nearly a century after he published those papers, mathematicians started taking interest in them internationally.

The second group consists of research papers on the Theory of Function of a Real Variable, the thrust being mainly on Fourier series. This was the area of Professor Prasad's main interest. The work on this area constitutes the bulk of his research. He has written out a number of results concerning the summability and strong summability of Fourier series. Most of his work in this field is based on a special type of functions having discontinuities of the second kind.

In his paper titled *'On the present state of the theory and application of Fourier Series'* [Bulletin of the Calcutta Mathematical Society, Vol. 2, (1910–1911), 17–24], Ganesh Prasad in his concluding remarks posed three problems mentioned as (A), (B) and (C) before the mathematicians of the day. The mathematical expressions being too cumbersome are not being mentioned here. However, the historical development connected to these is interesting. Incidentally, it may be mentioned that problems proposed by Professor Ganesh Prasad are fundamental to the theory of Fourier series. Later a Russian mathematician named N. N. Luzin (1883–1950) conjectured about the second problem (B) as set by Professor Prasad. In short, his conjecture was that *'the Fourier series of a square integrable function including continuous functions in particular converge almost everywhere'*. In 1966, a famous Swedish mathematician Lennart Carleson (b. 1928) proved this conjecture to be correct. In 2006 he won the Abel Prize for this major contribution.

The above incident highlights the fact that Ganesh Prasad was very much aware of the fundamental problems of the subject he was dealing with. Through his own research papers, he drew the attention of the international community of mathematicians to these challenges. That is how the problem posed by him in his publication of 1910 was ultimately solved by L. Carleson in 1966 and fetched him the Abel Prize.

He also delivered a series of six lectures on the researches in the Theory of Fourier series. In 1928, the University of Calcutta published these lectures in the form of a book entitled '*Six lectures on Recent Researches in the Theory of Fourier series*' [University of Calcutta (1928)]. In this book, he has discussed in details the convergence problem of Fourier series. It may be noted that Ganesh Prasad made a notable publication titled '*On the summability (C1) of the Fourier Series of a function at a point where the function has an infinite discontinuity of the second kind*' [Bulletin of the Calcutta Mathematical Society, Vol. 19, pp. 51–58].

His third set of research papers are on Spherical Harmonics and Legendre's function. It may be noted that the Spherical Harmonics are the angular portion of the solution to **Laplace's equation** in **spherical coordinates** where azimuthal symmetry is not present. He examined the expansion of functions in a series of Spherical Harmonics.

Prof. Prasad has given an interesting theorem using which an arbitrary function can easily be expressed as a sum of surface harmonics [Vide Hobson, '*Theory of Spherical and Ellipsoidal Harmonics*'; University of Cambridge, (1931) p. 148]. This book written by E. W. Hobson was later again reprinted in 1955. In the book mentioned above, the writer referred to the work of Dougall, published in 1913 and at the same time also commented that a similar work had already been published by Ganesh Prasad in 1912. In this context, Hobson wrote:

> A somewhat different, but equivalent, form for f_n (polynomial function of degree n) had already been given by G. Prasad. By this formula the value of f_n over a sphere $r = a$ is expressed as a sum of surface harmonics.

In this context it may be mentioned that Prof. Ganesh Prasad's book entitled '*A Treatise on Spherical Harmonics and the Functions of Bessel and Lamé*' is treated as a classic work in Mathematics.

The most remarkable fact about Prof. Ganesh Prasad's research work was that he always took up challenging problems in areas which were very relevant in those times. He himself worked in such areas and motivated his research students to further investigate the associated problems.

He was a remarkable teacher and research guide in modern higher Mathematics in colonial India. That is why, the mathematician Vice-Chancellor of Calcutta University, Sir Asutosh Mookerjee invited and persuaded him twice to take up the responsibilities of both the Mixed and Pure Mathematics Departments of his University. Professor Ganesh Prasad was successful both the times, and he was responsible in building up strong schools of research in both the Departments much to the satisfaction of Sir Asutosh.

To elucidate the point, we shall take up a case or two. For example, in the context of his paper entitled '*On the fundamental theorem of the Integral Calculus for Lebesgue Integrals*' [Bulletin of the Calcutta Mathematical Society, 16, (1926) 109–116.], the following features are noteworthy. In two of his earlier papers namely '*On the fundamental theorem of Integral Calculus*' [Bulletin of the Calcutta Mathematical Society, 15, (1924–25), 1–4.] and '*On the fundamental theorem of the Integral Calculus in the case of Repeated Integrals*' [Bulletin of the Calcutta Mathematical Society, 16, (1926) (1–8)] he had studied the problems of:

Deciding (1) if the differential coefficient of the Riemann indefinite integral of a function having a discontinuity of the second kind at a point in the region of integration, exists at that point or not, and (2) finding the value in the former event, that is if it exists.

After these calculations, his objective in the third paper was to show, by selected examples, that for the Lebesgue integral of a function, whose discontinuities were of the second kind, many features present themselves during the course of the solution of the problem which were similar to those found in the papers referred to earlier, although the function was not integrable according to Riemann's definition.

Historically, it may be noted that the first serious contribution to the study of the problem was the paper entitled '*Uber die differenzirbarkeit eines Integrales nach der oberen Grenze*' [Nachrichten d Königlichen Gesellschaft der Wisseschaften zu Göttingen, (1893), pp. 696–700]. The paper was communicated by late Professor J. Thomae of Jena to the Royal Society of Göttingen in October, 1893. The introduction to the paper, translated into English would be: 'If $f(x)$ is a function integrable between the limits a and b if for a value c of x between a and b, $f(c + 0)$ does not exist', then the question that arises is whether, $\omega(x) = \int_a^x f(x)\mathrm{d}x$, possesses a progressive differential coefficient at the point C or not. Since, in general, the progressive differential coefficient of $\omega(x)$ is equal to $f(x + 0)$, it is natural to conclude that this differential coefficient does not exist at the point C. Prof. Thomae proved the above-mentioned theorem by using Du Bois-Reymond's Second Mean Value theorem.

In the well-known book entitled '*Bericht Uben die Entwickelung der Lehre Von der Punktmannigfaltigkeiten*' (Leipzig, 1900), Prof. A. Schoenflies devotes a whole chapter to the fundamental theorem of the Integral Calculus.

Professor Ganesh Prasad further suggested the following problem: '*To classify non-differentiable functions according to the number of limiting points which its zeros possess in a finite interval*'.

S. K. Bhar was encouraged by Ganesh Prasad to carry out the investigation.

Professor Ganesh Prasad's own research scholar Dr. Avadesh Narayan Singh had published two original research papers in this area under his guidance.

In the first paper Dr. Singh had defined a class of Non-differentiable functions which was published in the 'Annals of Mathematics' [Vol. 28, (1927), pp. 472–476].

In the second paper of the series entitled '*On the unenumerable zeros of Singh's of Non-differentiable functions*' [Bulletin of the Calcutta Mathematical Society, 22, (1929), 91–102], Dr. A. N. Singh has studied the zeros of this class of functions and made useful conclusions.

Taking up a third example, we shall once again show the mastery of Dr. Ganesh Prasad as a research guide. As early as in 1915, Prof. Ganesh Prasad published a research paper entitled '*On the existence of the mean differential coefficient of a continuous function*' [Bulletin of the Calcutta Mathematical Society, 3, (1911–12), 53–54]. In this paper, Prof. Prasad raised a question and asked *'Does there exist a continuous function which has no mean differential coefficient for any value of x?'*.

Encouraged by Ganesh Prasad, one of his students Santosh Kumar Bhar took up the problem. Prof. Prasad in his own paper, had investigated the non-differentiable functions of Weierstrass, Darboux, Lerch, Fabex and Landsberg. He found that each of these functions has a set of points—everywhere dense, where the mean differential coefficients exist. S. K. Bhar carefully examined some of the non-differentiable functions, which had been discovered at that time, such as those of Steintz [Mathematische Annalen, Bd. 52 (1892), pp. 58–62], Peano [Mathematische Annalen, Bd. 26], A. N. Singh [Annals of Mathematics, 28 (1927), 472–476] and Hahn [Jahresber d.d Math Vereinigung, Bd. 26 (1918), pp.281–284] and came to identical conclusions. Thereafter S. K. Bhar took up some non-differentiable functions not investigated by Ganesh Prasad and ultimately discovered some important results in the case of Dini's function [Annali di Mathematica, Ser. 2, t, (1877), 8].

S. K. Bhar concluded that.

I.   Generally, the mean differential coefficient is non-existent at every rational point in $(0, 1)$.

II. It is completely proved that in all cases of irrational numbers, the mean differential coefficient is non-existent.

Thus encouraged and inspired by Dr. Ganesh Prasad, S. K. Bhar published a notable research paper named *'On a continuous function which has no mean differential coefficient for any value of x'*, which was published in the Bulletin of the Calcutta Mathematical Society.

Prof. Ganesh Prasad had independently suggested the problem to one of his students Dr. Lakshmi Narayan. Without any previous knowledge of Thomae's results, L. Narayan built up a theory and called it 'integration—image'. However, Prof. Ganesh Prasad thought that the method of Dr. Lakshmi Narayan was not suitable for giving a general criterion. While going through the manuscript of Dr. L. Narayan's paper [which was published in the 'Proceedings of the Banaras Mathematical Society' in the years 1924–1925], Prof. Ganesh Prasad became interested in the problem. He worked on all the three connected problems mentioned at the outset. The criteria given by him are believed to be simpler and more general than the criterion given by Thomae.

Mr. J. M. Whittaker, a well-known British mathematician also became interested in the problem after reading Dr. Lakshmi Narayan's work. Later he published a paper entitled *'The differentiation of an Indefinite Integral'* [Proceedings of the Edinburgh Mathematical Soc., Vol. 46, part 2, Nov. 1925].

This discussion clearly indicates Prof. Prasad's exceptional talent in solving difficult mathematical problems and also his capacity to motivate other mathematicians to take up the associated unsolved problems.

Another research paper written by Professor Ganesh Prasad entitled '*On the zeros of Weierstrass's non-differentiable function*' ('Proceedings of the Banaras Mathematical Society', Vol. XI, 1930, pp. 1–8) created lot of interest among many mathematicians of those times. Professor Prasad inspired many of his students to work in that area and at least three very good papers were published by his students. To exhibit Professor Prasad's remarkable ability to guide, encourage and draw out the best from his students, we would like to cite some of the remarks made by the involved mathematicians.

Encouraged and helped by Prof. Ganesh Prasad, Prof. Bholanath Mukhopadhyay published a paper entitled '*On the limiting points of the zeros of a non-differentiable function first given by Dini*' (Bulletin of the Calcutta Mathematical Society, Vol. XXII, No. 2 and 3, 1930, pp. 103–114), and in the introduction to his paper he writes:

> *The publication of Dr. Ganesh Prasad's remarkable paper 'On zeros of Weierstrass's Non-differentiable functions' has naturally led mathematicians to study the zeros of different types of Non-differentiable functions.*

Another mathematician S. K. Bhar published a paper entitled '*On the zeros of Non-differentiable functions of Darboux's type*'.

S. K. Bhar in the course of the paper wrote:

> *.....those who wished to give graphical representation of non-differentiable functions, as Weiner, Felix Klein and G. C. Young did in the case of Weirstrass's function, must have desired to know the zeros of the concerned function. However, it is only recently, that the first successful investigation of this type has been published by Dr. G. Prasad, who has given general expressions from which zeros of Weirstrass's function can be obtained.*

His remarkable ability to inspire students produced a rich harvest of brilliant research papers in mathematics. Prof. Prasad inspired N. N. Ghosh and he wrote and published a paper entitled '*On the calculation of the zeros of Legendre Polynomials*' [Bulletin of the Calcutta Mathematical Society, 21 (1) (1929), 61–68].

Prof. G. Prasad encouraged and guided H. P. Banerjee to publish a paper entitled '*On the summability (C, 1) of Legendre series of a function at a point where the function has a discontinuity of the second kind*'.

The above examples demonstrate how very successfully, Ganesh Prasad motivated his students to take up challenging and important mathematical problems and solve them. In this context, it may be kept in mind that in the British-ruled colonial India of the nineteenth century funds were very meagre in teaching and research. But in spite of that Prof. Ganesh Prasad successfully attracted many talented students to choose and devote their life in such underpaid vocations.

Apart from being a great teacher, a great researcher and a brilliant research guide in different branches of Pure and Applied Mathematics, Prof. Prasad also wrote a number of articles on 'History of Mathematics' and published a well-known book on 'Some great Mathematicians of the Nineteenth Century: Their lives and works'. In this treatise he has discussed about 16 mathematical greats such as A. L. Cauchy (1789–1857), C. F. Gauss (1777–1855), C. G. J. Jacobi (1804–1851), N. H. Abel (1802–1829), K. T. W. Weierstrass (1815–1897), G. F. L. P. Cantor

(1845–1918), J. H. Poincare (1854–1912), G. F. B. Riemann (1826–1866), J. G. Darboux (1842–1917), M. G. Mittag-Leffler (1846–1927), C. F. Klein (1849–1925), C. Hermite (1822–1901), F. Brioschi (1824–1897), L. Kronecker (1823–1891), A. L. G. G. Cremona (1830–1903) and A. Cayley (1821–1895) and elaborated on their mathematical achievements and contributions.

In this area also, he strongly influenced and inspired two of his students in Calcutta University—named B. B. Datta and A.N. Singh. He inspired them to carry out serious research in the fields of ancient Hindu and Jaina mathematics.

Dr. Prasad had published about 35 original research papers in national and international journals and 8 books. He attracted students from different parts of India, who were eager to carry out research under his guidance. Some of his distinguished students are Professor B. N. Prasad of the Allahabad University, Professor A. N. Singh of the Lucknow University, Professor Braj Mohan of the Banaras Hindu University and Professor Hariprasanna Banerjee of the Calcutta University. A list of his well-known books is given below:

- *Differential Calculus (1909),*
- *Integral Calculus (1910),*
- *An Introduction to Elliptic Function (1928),*
- *A Treatise on Spherical Harmonics and the Functions of Bessel and Lame (1930–1932),*
- *Six Lectures on recent researches in the theories of Fourier Series (1928),*
- *Six Lectures on recent researches about Mean-Value Theorem in Differential Calculus*
- *Mathematical Physics and Differential Equations at the beginning of the twentieth Century*
- *Some great mathematicians of the twentieth Century: Their lives and works. (2 Volumes).*

He kept himself informed about the contemporary research work being carried out on the branches of Mathematics, which interested him in different European countries. He also wanted his students to be aware of the new developments that took place in Mathematics and encouraged them to take up researches on that line. His special lectures delivered from time to time at different places was full of such materials. His inaugural lecture after joining as the Rash Behary Ghosh Professor at the Department of Mixed Mathematics of the Calcutta University in 1914 was titled '*From Fourier to Poincaré: A Century of Progress in Applied Mathematics*' reflected the depth of his knowledge and understanding of the subject.

This erudite scholar, brilliant mathematician and popular research guide passed away at Agra on the 9 March 1935.

4.3 Important Milestones in the Life of Professor Ganesh Prasad

1876: Born on the 15 November 1876, at a small town Ballia in U. P.

1891: Passed the Matriculation examination in first division.

1895: Graduated from Muir Central College, Allahabad with a First class in Mathematics (Honours).

1897: Obtained M.A. degrees from the Universities of Allahabad and Calcutta.

1898: Received the D.Sc. degree from the University of Allahabad.

1899: Proceeds to England with a Government of India Scholarship for pursuing higher studies in the Cambridge University. Subsequently he moved over to Germany to work under the guidance of Professor F. Klein.

1904: Returns to India.

1905: Appointed Professor of Mathematics at Queens' College, Banaras.

1914: Joins the University of Calcutta as the first Rashbehary Ghosh Professor in the Department of Mixed Mathematics (Later called the Department of Applied Mathematics).

1917: Leaves the Calcutta University and joins the 'Banaras Hindu University' as Professor of Mathematics.

1923: Returns to Calcutta and joins the Calcutta University again as the first Indian 'Harding Professor' in the Department of Pure Mathematics.

1935: Elected Foundation Fellow of the Indian National Science Academy.

1935: Passed away at Agra on 9 March 1935.

Chapter 5
Bibhuti Bhusan Datta (1888–1958)

5.1 Birth, Family, Education

Bibhuti Bhusan Datta was born on the 28 June 1888, in Kanungopada village of Chittagong in undivided Bengal [present-day Bangladesh]. His father Rashikchandra Datta was employed in a subordinate post in the judicial department. His mother Muktakeshi Devi was a religious and kind lady. His parents though poor were honest, religious and very helpful people. Bibhuti Bhusan was the third of the couple's eleven children. Right from his childhood, Bibhuti Bhusan showed signs of great intelligence.

A close observation of the life of Bibhuti Bhusan Datta clearly indicates two distinct traits in his persona. In the initial years of his life he was a mathematician, a famous teacher of Mathematics at the University of Calcutta, a highly intelligent mathematical investigator and a pioneer researcher in 'History of Mathematics' committed to reinstate the glory of ancient Hindu Mathematics. In his later years, he became a saint and religious philosopher who renounced all worldly possessions and comforts. In the present write-up the main emphasis is in highlighting his role as a pioneer mathematician, who enriched the mathematical scenario of Bengal as well as India with his original research in Applied Mathematics and his pioneering research in the history of ancient Hindu Mathematics.

B. B. Datta passed his Entrance Examination in 1907 from Chittagong Municipal School in first division with a merit scholarship. Then he came to Calcutta and joined the famous Presidency College to pursue further studies. In 1909, he passed the Intermediate Examination in Science (I.Sc.) with a first division but did not get a merit scholarship. In 1912, he graduated with Honours in Mathematics from the Scottish Church College, Calcutta. After graduation, B. B. Datta joined the newly established Department of Mixed Mathematics (later known as Applied Mathematics), University College of Science, Calcutta University. In 1914, he passed the M.Sc. Examination in Mixed Mathematics from the University of Calcutta and in the year 1915, he obtained the Rashbehari Ghosh scholarship and started research work on Hydrodynamics.

P. Mukherji, *Notable Modern Indian Mathematicians and Statisticians*,
https://doi.org/10.1007/978-981-19-6132-8_5

5.2 Career, Research Contributions

In 1916, B. B. Datta was recruited to assist the then Rashbehari Ghosh Professor Dr. Ganesh Prasad. In 1917, he was appointed as a lecturer in the Department of Applied Mathematics, University of Calcutta. It would be pertinent in this context to point out that the then Vice-Chancellor Sir Asutosh Mookerjee, the great connoisseur of academic excellence was the man behind this appointment. He had once again rightly chosen a brilliant mathematician to serve the Department of Applied Mathematics. Thus B. B. Datta started his teaching career and was considered to be a very good teacher by the students of the department. He was equally at ease teaching Statics, Dynamics, Hydrostatics, Astronomy and even a new subject for those times such as Theory of Planets. Along with his teaching B. B. Datta concentrated on his research activities too. In 1919, he obtained the prestigious Premchand Roychand studentship of the University of Calcutta and was awarded the Mouat Medal and the Elliot prize for his original contributions in the field of Applied Mathematics. In 1921, he obtained the D.Sc degree from the University of Calcutta for his outstanding research work in the field of Hydrodynamics. Some of his important research contributions in various areas of Applied Mathematics published in noted national and international journals are listed below:

(1) '*On a method for determining the non stationary state of heat in an ellipsoid*' [American Journal of Mathematics, Vol. 41, 1919, pp. 133–142],
(2) '*On the distribution of electricity in the two mutually influencing spheroidal conductor*' [Tohoku Mathematics Journal, 1920, pp. 261–267],
(3) '*On the stability of the rectilinear vortices of compressible fluids in an incompressible fluid*' [Philosophical Magazine, Vol. 40, 1920, pp. 138–148],
(4) '*Notes on vortices of a compressible fluid*' [Proceedings of the Banaras Mathematical Society, Vol. 2, 1920, pp. 1–9],
(5) '*On the stability of two co-axial rectilinear vortices of compressible fluids*' [Bulletin of the Calcutta Mathematical Society, Vol. 10 (No. 4), 1920, pp. 219–220],
(6) '*On the periods of vibrations of straight vortex pair*' [Proceedings of the Benaras Mathematical Society, Vol. 3, 1921, pp. 13–24],
(7) '*On the motion of two spheroids in an infinite liquid along their common axis of revolution*' [American Journal of Mathematics, Vol. 43, 1921, pp. 134–142],

Some of his research work on the following topics remained unpublished:

(1) On the motion of spheroids in infinite fluids.
(2) Development of the perturbative function when the

    (i) eccentricities and (ii) the mutual inclinations of the orbits are small.

(3) On the conduction of heat in an ellipsoid with three unequal axes.
(4) On the non-stationary state of Heat in an Ellipsoid.
(5) On the motion of two spheroids in an infinite liquid along the common axis of revolution.

(6)   Inequalities arising from figure of the earth.
(7)   On the conduction of heat in an ellipsoid of revolution.
(8)   Secular variations.
(9)   On the transformation theorem relating to spheroidal harmonics.
(10) On the transformation theorem relating to spheroids.

According to experts these unpublished research works were of considerable merit.

Bibhuti Bhusan Datta was very keen to relate fundamental theories to their actual physical interpretation.

As an example, we may mention, that in one such paper '*On a physical interpretation of certain formulae in the Theory of Elasticity*', B. B. Datta showed how certain formulae relating to the theory of elasticity of a homogeneous medium may be interpreted in a manner so as to lead to an explanation of Gravitation.

Another research paper entitled '*On the Figures of Equilibrium of a Rotating Mass of Liquid for Laws of Attractions other than the Law of Inverse Square*'—Part I which he took up for investigation under the guidance of Dr. Ganesh Prasad, was also quite interesting. The paper contains the results of his findings about the figures of equilibrium of a rotating mass of a homogeneous incompressible liquid whose particles attract one another according to the laws of force other than the Newton's law of Gravitation.

In the third part of this same paper, B. B. Datta proved a theorem for the law of direct distance which is analogous to a theorem of Poincare's for the law of Inverse Square of the distance.

Clearly, B. B. Datta had a penchant for generalizing the results of famous mathematicians and their propounded theories. His research in Applied Mathematics had a lot of scope for application to physical and real-life problems. Two research papers of B. B. Datta which have relevance in applicability to practical problems even in the contemporary scenario are being mentioned here. In the early part of the twentieth century, he considered the problem of determining the non-stationary state of heat in an ellipsoid. While solving this problem, he generalized the equation of the elliptic surface using solid harmonics. Then using boundary conditions, he obtained an algebraic equation. He also found out a mathematical method for obtaining approximate values of the roots of that equation. He also calculated the changes in the thermal field in a direction perpendicular to the elliptic plane. In solving this problem, he obtained two transcendental equations. He completed this research paper in 1917, and it was finally published in 1919 as a research paper entitled '*On a method for determining the non-stationary state of heat in an ellipsoid*' [American Journal of Mathematics, Vol. 41, 1919, pp. 133–142],

Incidentally, it may be noted that with the discovery of antennas and other electrical devices, in the recent times, the solution of physical problems related to elliptical and spherical surfaces has become very important.

During the early part of the twentieth century, research in Mathematics in our country was in its infancy. No research guides were available. There were no hydrodynamical laboratories anywhere in India. So it was extremely difficult or to be more precise, almost impossible to carry out any experimental research work in Applied Mathematics and particularly in Fluid Mechanics. But in that situation, it is really amazing, how B. B. Datta worked on the problem of the stability of circular vortices. He solved the problem using mathematical techniques. He however had no means of verifying it experimentally. But, interestingly enough this problem is actually a real-life problem, which is frequently encountered during power generation by turbines. B. B. Datta also worked extensively on compressible fluids.

In the backdrop of the early twentieth-century mathematical world of India, B. B. Datta's research and contributions in Applied Mathematics are really awesome.

If B. B. Datta's research contributions, in the area of Fluid Mechanics is rich, then his contributions in the field of History of Mathematics is stupendous. India, as is well known had from ancient times, a rich tradition in Mathematics. Many great mathematicians were born here and made fundamental work in the fields of Arithmetic, Algebra, Geometry and Trigonometry. But unfortunately, very few well-documented records were available. In the nineteenth and earlier part of twentieth century, mathematicians from the Western world projected 'Hindu Mathematics' as a part of astrology. They proclaimed that Hindu Mathematics was a tool for making astrological predictions. With relentless perseverance, and indomitable efforts, B. B. Datta was the first mathematician who expressed the ancient Indian Mathematics as a part of internationally acclaimed 'Pure Mathematics'. He very successfully demonstrated that the ancient Hindus pursued research in Pure Mathematics and latter applied it to practical problems. He became famous for his original research in the history of ancient Hindu Mathematics.

In 1923 the then Principal of Varanasi Central College and eminent mathematician Prof. Ganesh Prasad was brought back to the Calcutta University for the second time as the 'Hardinge Professor of Higher Mathematics' in the Department of Pure Mathematics. This was a turning point in the life of Dr. B. B. Datta. Prof. Ganesh Prasad himself was a famous historian of Mathematics. He inspired his two students Bibhuti Bhusan Datta and Avdesh Narayan Singh to take up research in the field of ancient Hindu Mathematics seriously. Dr. B. B. Datta was a great patriot. He greatly admired and respected the ancient Hindu and Jain mathematicians and philosophers. He was already deeply hurt by the comments and attitude of many historians of mathematics, who without going deeply into the subject, boldly declared about the indebtedness of Hindu mathematicians to external sources. There was also a tendency among many foreign historians of Mathematics to totally ignore the discoveries made by ancient Indian mathematicians and bestow credit to later day mathematicians of other countries. Sometimes by wrong analogy and interpretation, they refused to accept the contributions of Hindu mathematicians.

Dr. B. B. Datta was greatly disturbed by such events. Prof. Ganesh Prasad's influence on him acted like a catalytic agent, and around that time, he finally gave up his researches in Applied Mathematics and whole-heartedly devoted himself in unravelling the history of Hindu Mathematics. But the path was difficult. By now it is widely known that many of the rich discoveries and contributions in science made by ancient Hindu scholars are embodied in the ancient Sanskrit books written by renowned sages. The tradition of the ancient sages was to compose everything in poetic form. Even the mathematicians followed this custom and their books consisted of a large number of Sanskrit Slokas. For writing in that poetic fashion, one had to follow the rules of rhymes and metres. Though there were many technical terms involved and the ancient Hindus knew word numerals, still for expressing the mathematical discoveries in poetic pattern alternative suitable words were found and freely used. Moreover because of the absence of printing press, the original writings of the masters were copied by the students and there could have been some errors in them. With the passage of time, it was difficult to identify the original. As Sailesh Das Gupta, a scholar on B. B. Datta has pointed out: 'The greatest advantage in this respect was the copious number of commentaries of such books which direct one to the true course. More than 100 commentaries have been written on Bhaskaracharya's Lilavati.

It will be the duty of the researcher to understand the objective of the writer. Though each word may have more than one meaning, he must catch the right content by analysing it in the proper context.'

[Ref: Mathematician Bibhuti Bhusan Datta–Sailesh Das Gupta, Read on 28-06-1988 on behalf of Calcutta Mathematical Society].

These were the immense difficulties involved, in the way of bringing to light the contributions of ancient Hindu and Jaina mathematicians in its proper perspective. A very large section of even the Indian population was unaware of these great achievements by ancient Hindu and Jaina mathematicians and because of wrong interpretation of the content, wrong deductions were made, catering incorrect information to the people.

Dr. B. B. Datta, however achieved immense success in his venture. He was truly a great mathematician. So with a logical mind he attempted to decipher the writings of the ancient mathematicians. In this context, it is noteworthy that the famous historian of Mathematics, Mr. D E Smith very challengingly stated: 'As to the spread of the Hindu-Arabic Numerals in Europe, as to the origin of the numerals, as to the origin of the symbol zero, as to the early history of Roman numerals and as to the place value in notation—such problems have long attracted the attention of scholars, but fields have by no means been fully achieved.'

For Dr. B. B. Datta this must have been a major challenge, which he took up and answered practically all the queries through his various articles and books. Dr. Datta while establishing his own theory or while contradicting a theory against the contribution of Hindu mathematicians first mastered all the facts about that particular theory. These included points both in favour of the theory as well as those against it. Then he critically analysed, rejected some and accepted others. On this basis of what he had accepted, he drew a conclusion.

Dr. B. B. Datta wrote nearly 60 articles on contributions of ancient Hindu and Jaina mathematicians. Among these, some of the articles were published in reputed foreign journals like 'American Mathematical Monthly', 'Bulletin of the American Mathematical Society', 'Quellen und Studien Zur Geschichte der Mathematik', 'Archeion' and 'Scientica'. Many of these articles were written to correct the wrong accusations and wrong deductions related to the contribution of Hindu mathematicians.

Some relevant portions from his various articles are being quoted here to elucidate the point. In an article entitled '*Geometry—Hindu Origin of Geometry*', Dr. B. B. Datta writes:

> The Hindu Geometry commenced at a very early period, certainly not later than that in Egypt, probably earlier, in construction of alters for the Vedic sacrifice. In course of time it however, grew beyond its original sacrificial purpose or the limits of practical utility and began to be cultivated as a science for its own sake also. Indeed, there is no doubt, about the fact that the study of geometry as a science began first in India. Further, the early Hindu Geometry was much in advance of the contemporary Egyptian or Chinese Geometry. The Greek Geometry was yet to be born.

In this connection a remark by the noted historian R C Dutta ['Short History of Greek Mathematics',—Journal of Gower Society, Cambridge, 1884, p 129], is very relevant. He writes: 'Though the Hindu achievements in Geometry were overshadowed at a later period, it would never be forgotten that the world owes its first lessons in Geometry not to Greece, but to India.'

Dr. B. B. Datta in a very scientific way established these facts. Just to elaborate the point, an example is cited. Greek Geometrician Democritus (400 B.C) referred to the Egyptian geometers by the term 'Harpedonoptae' ['A history of Civilization in Ancient India'—Romesh Chandra Dutta, Vol. 1, 1893, pp. 70]. The famous mathematician—Cantor (1845–1918) explained that the word was created by compounding two Greek words and means 'rope-fastener' or 'rope-stretcher'. In this context, Dr. B. B. Datta rightly reasoned that in Greece, as well as in Egypt, geometry was always referred to as 'land-measurer'. So the conception of the word 'rope-fastener' or 'rope-stretcher' is neither Greek nor Egyptian. Whereas in Sanskrit the synonym of geometer is 'śulba-bid' which when translated mean 'the expert in rope-measurement'. Actually the ancient Hindu treatises on geometry are called 'Sulba-Sutras' or the 'rules of rope measurement'. On the strength of such findings B. B. Datta commented: 'Hence Democritas, I presume, got his term for the geometers in his Indian travels. Here is then probably a specific instance of the contact of the Hindu and Greek geometry which almost all writers believe more or less'.[B. B. Datta: '*Origin and History of the Hindu names for Geometry*'; Quellen und Studien, B. 1, Heft 2, 1929].

In the same article, B. B. Datta has commented that 'The use of the word rajjŭ at least in its ordinary signification of rope occurs as early as Veda' [Rig—Veda, Atharva Veda, Satapatha Brahaman].

In a well-researched study on Algebra Dr. B. B. Datta established with conviction that the Arabs learnt about the subject from the Hindus. In this article on 'Origin of Algebra', Dr. Datta writes:

'The science of Algebra derived its name from the title of a certain work by the Arab Mohammad bin Musa, Alkhowarizmi (820 A.D) 'al-gebr-w'al mugābalah', which contains an early systematic treatment of the general subject as distinct from the science of numbers. That........ This book was translated by early western scholars together with other Arabic works on Algebra. So Europe first learnt of Algebra from the Arabs. But the subject matter of Alkhowarizmi's treatise was not his original contribution. He got it from the Hindus. Indeed as in arithmetic, so also in algebra, the Arabs got their first lessons from the Hindus.....................

The Hindus were indeed far in advance of all the other nations in analysis. Further, it should be remembered that the form and spirit of modern algebra are essentially Hindu, not Grecian, nor Chinese.

The Hindu contributions to Algebra are of fundamental character. Even some of the technical terms which are commonly used in modern algebra are of Hindu origin.'

Again in the context of use of calculus in Hindu Mathematics, Prof. B. B. Datta has written that: 'the use of a formula involving differentials in the works of ancient Hindu mathematicians has been established beyond doubt. That the notions of instantaneous variation and that of motion entered into the Hindu idea of differentials as found in works of Manjula, Āryabhatta II and Bhāskara II is apparent from the epithet Tatkalika (instantaneous) gati (motion) to denote these differentials'.

In another context, Dr. B. B. Datta wrote: 'The calculation of eclipses is one of the most important problems of Astronomy. In ancient days this problem was probably more important than it is now, because the exact time and duration of the eclipse could not be foretold on account of lack of the necessary mathematical equipment on the part of the astronomer. In India, Hindus observed fast and performed various other religious rites on the occasion of eclipses. Thus the calculation was of national importance. It afforded the Hindu astronomer a means of demonstrating the accuracy of his science.'

Between 1925 and 1947, Dr. Datta wrote at least 55 research papers, all concerning the Mathematics of the ancient Hindus and Jainas. Dr. Datta was very well conversant in Sanskrit. He had intensively studied the Vedas, the Upanishad, and the Puranas, the Sanskrit literature, the treatises on ancient Indian Philosophy and the ancient books on Mathematics written in Sanskrit. From these precious sources, B. B. Datta gathered thousands of factual materials and references. In 1935, Dr. Datta's classical book 'The History of Hindu Mathematics' Part I and II was published. Dr. A. N. Singh of Allahabad University was the co-author of the book. The two volumes of the book were first published from Lahore in 1935 and 1938 respectively. In Part I of this classic the history of Hindu numerals including decimal, place value system, together with other systems is clearly written. Notable historians of Mathematics like Rosen, Reinaud, Woepcke, Strachey, Taylor and many others did believe that the so-called Hindu-Arabic numeral system was the discovery of Hindus alone. But there remained some controversy in this matter. There were some historians of Mathematics who did not accept this view. Dr. B. B. Datta took up the challenge and collected thousands of factual materials and references and with the help of those, in this book he has proved the authorship of the present numerals, on a sound scientific basis. Thus from the so-called Hindu-Arabic numerals, the word 'Arabic' had to be dropped. The man responsible for this memorable task is none other than Dr. B. B. Datta.

Part II of the book deals with the contributions of Hindus in Algebra. It establishes the fact that Hindus had a remarkable measure of success in obtaining the general solution in rational integers of the indeterminate equation, biquadratic equation and barga prakanti (Pellian Equation). There was a third part of the book which remained unpublished during B. B. Datta's life time. After some investigation, the manuscript was traced. This part consists of Hindu Geometry, the contributions of Hindus in Trigonometry, Calculus, Permutation and Combination, Surds, Series and Magic Square. Out of these the History of Hindu Geometry, Trigonometry and Calculus were published in the 'Indian Journal of History of Science' (IJHS) during 1980–1983.

Now a brief account is recorded about Dr. B. B. Datta's other trait in his persona. Right from his childhood, B. B. Datta had a great fascination for asceticism. He preferred reading more books on Philosophy than on any other subject. A very simple man with a frugal life-style, he had no attraction for worldly possessions and his ultimate aim was to become a 'Sanyasi'. Although he was a much respected and admired Professor in the Department of Applied Mathematics, from 1928 onwards, he became very irregular in attending the University. Around 1930, he actually tendered his resignation from the Department of Applied Mathematics and moved on to a famous pilgrimage centre named Pushkar in Rajasthan and decided to stay there permanently. But his teacher Prof. Ganesh Prasad, who was then the Hardinge Professor of Pure Mathematics, persuaded Dr. Datta to return to the University of Calcutta. Through the efforts of Prof. Ganesh Prasad in 1931, Dr. B. B. Datta was appointed as a special 'Readership Lecturer'. He was allowed to conduct only research work and was relieved of teaching duties. In 1931, Dr. B. B. Datta delivered six lectures about ancient Hindu Geometry. In 1932, the University of Calcutta published these lectures in the form of a book. The book was titled *'The Science of Sulba—A Study of Hindu Geometry'*. This book is still very much in demand worldwide among researchers of ancient Indian Mathematics.

These lectures were the fruit of intensive research by B. B. Datta. He revealed that the Vedic priests were concerned with the practical problems involved in the construction of altars for carrying out religious rituals. In this way they developed a type of 'esoteric geometry as their secret property.'

Dr. Datta found out that the problems that arose at that time for constructing altars of different shapes with given areas and fixed number of bricks needed 'a varied knowledge about properties of geometrical figures including their areas, idea of mensuration, knowledge of arithmetic and even algebra were essential for due performance'.

The mathematical problems that naturally arose were:

- Division of any figure in any assigned number of parts.
- To draw a straight line at right angles to another straight line, from an external point.
- To construct the geometrical shapes adhering to defined areas.
- Combination and Transformation of areas.
- Use of the formula $x^2 + y^2 = z^2$ in connection with Pythagoras's theorem and finding out 3 suitable numbers satisfying the equation both for integers and rational numbers.
- Similar figures having different areas. For carrying out related calculations the problem of finding the value of $\sqrt{2}$ became necessary. Greek mathematicians also encountered similar problems. But they left it as unsolved. But this irrational

number did not trouble the ancient Hindu mathematicians. In fact the Hindu mathematicians found out very good approximations to the value of $\sqrt{2}$ as well another irrational number π. Baudhayana and Apastamba's formula gave the approximate value of $\sqrt{2} = 1.4142156$ which is correct to 5 decimal places. Aryabhata gave a very close approximate value for π which is equal to 3.1416. This was the best value determined till that time.

But many historians of Mathematics were not prepared to accept such high degree of accuracy from Hindu mathematicians. Rodet, Smith and Heath attributed it to Greek influence. B. B. Datta after intense research in this area rejected the hypothesis of the foreign historians on three counts.

He stated that:

1 Such accuracy was unknown to Greek mathematicians of that time.
2 The Hindu and Jaina mathematicians had been using the number 'ayuta' long before the Greeks became aware of such a number word of high denomination.
3 Aryabhata's result was achieved using Trigonometry. The Greeks did not know Trigonometry at time.

Dr. Datta did painstaking research to establish and prove before the world the supremacy of ancient Hindu Mathematics. Once convinced about a fact Dr. Datta was very bold in defending his findings. His research on 'Alphabetic numeral system' is a case in point. In the related article, published in 1929 Dr. Datta made a very elaborate and logically rigorous discussion. He stated that the custom of writing numerals with the help of alphabets was quite common among the Indians, the Greeks, the Jews and the Arabs. He discussed the method followed by ancient Indians in great details. According to Dr. Datta, famous Indian mathematician Aryabhat I used 42 alphabets for his method. He followed 'Shiva sutra'. Panini too had made use of 'Shiva sutra'. Dr. Datta in his research paper, has fearlessly pointed out the discrepancies in the related research work of J. J. Fleet. He has commented that probably Fleet was misled by the writings of Alberuni. Dr. Datta was also quite critical about the writings of Shankar Balakrishna Dikshit (Indian Astrology, 1876), Sudhakar Dwivedi (Ganaktarangini), Shri Gourishankar Hirachand Ojha (Ancient Indian Alphabets—2nd ed., 1918), C. M. Whees (On the Alphabetic Notation of the Hindus: Transactions of the Literary Society of Madras, part I, 1827). In 1935, E. Jacquet translated Dr. Datta's article in French and it was published in the same year in 'Journal Asiatique'. There was a lot of follow up publications to this article, by L. Rodet [Sur la Varitable Significance de la Notation numerique inventee per Aryabhatta, Journal Asiatique, 1880, Part II], M Cantor [Geschichte ter Mathematik, Bd 5, Leipzig, 1907] and G. R. Kaw [Notes on Indian Mathematics Arithmetical Notation—Journal of Asiatic Society of Bengal, Vol. 3, 1907].

In a Bengali article on 'Acharya Aryabhatta and his disciples', published in 1933, Dr. Datta discussed in details the books written by the great mathematician and drew his own conclusions about Aryabhatta's birthplace. He also gave some new information about Aryabhatta's disciples. According to Dr. B. B. Datta apart from Pandu Rangaswami, Latdev and Nishankuba Shanker, Prabhakar, Bhaskar and Lalla were also Aryabhata's students. The famous historian of Mathematics, Kripashankar Shukla has proved in one of his articles that Bhaskar was not a student of Aryabhata. He has rejected Dr. Datta's theory with adequate logical reasoning. Many historians of Mathematics have indicated South India as the birthplace of Aryabhata. Shankar Balakrishna Dikshit has commented that almanacs following Aryabhata's Siddhanta are used in South India and especially by the Vaishnavite Sects of Karnataka and Mysore. So Dikshit believes that Aryabhata was born in South India. Sambasiu Shastri has also supported this view. Dr. B. B. Datta with the help of beautiful logic has rejected these views and established that Aryabhata was born in a place called Kusumpur close to the modern city of Patna.

For a long time Dr. B. B. Datta had a slender relationship with his alma mater, the University of Calcutta. In 1933, he finally took the decision to retire from there at the age of 45. After that, he moved from place to place all over India. He was in acute financial distress and due to his ascetic life-style and inadequate nutrition, his health started deteriorating. He spent the last few years of his life under the adopted name 'Swami Vidyaranya' and stayed in an 'ashram' at Puskara in the state of Rajasthan. He lost his mother in March, 1958. Soon after that Dr. B. B. Datta bade the last good bye to this mortal world on the 6th October, 1958 and left for his much desired heavenly abode.

Dr. Bibhuti Bhusan Datta will be long remembered for the unique combination of mathematical brilliance and saintly leanings in his personality. His contributions as a researcher in the History of ancient Hindu Mathematics have earned him a permanent place among the historians of Mathematics in the world. His books and articles are indispensable source materials for any serious researcher in this discipline. In the annals of history of Mathematics in India, B. B. Datta will always be a name to be cherished and remembered with reverence and respect. It was pioneers like him who accounted for the golden age of Mathematics in Bengal during the twentieth century.

5.3 Important Milestones in the Life of Dr. Bibhuti Bhusan Datta

1888: Born on the 28 June 1888, in the district of Chittagong in erstwhile East Bengal.

1907: Passed the Entrance Examination in first division with a merit scholarship.

1909: Passed the Intermediate Examination in Science (I.Sc.) from Presidency College, Calcutta.

1912: Graduated with Mathematics (Honours) from the Scottish Church College, Calcutta.

1914: Passed the M. Sc. Examination in Mixed Mathematics from the Department of Mixed Mathematics, University College of Science, Calcutta University.

1917: Appointed a Lecturer in the Department of Mixed Mathematics, University of Calcutta.

1919: Obtained the Premchand Roychand Studentship (PRS) from Calcutta University.

1921: Awarded the D. Sc. Degree by the University of Calcutta for original contributions in Applied Mathematics.

1930: Tendered his resignation to the Department of Applied Mathematics and leaves to lead an ascetic life in Pushkar in Rajasthan.

1931: Returned to Calcutta University and joined as a special 'Readership Lecturer' there.

1932: Dr. Datta's classic book titled 'The Science of Sulba—A Study in Hindu Geometry' published by the University of Calcutta.

1933: Finally retired from the University of Calcutta at the age of 45.

1935: 'The History of Hindu Mathematics (Part I)'; authored by B. B. Datta and A. N. Singh published from Lahore [now in Pakistan].

1938: 'The History of Hindu Mathematics (Part II)'; authored by B. B. Datta and A. N. Singh published from Lahore [now in Pakistan].

1958: Breathed his last on the 6 October 1958, at Pushkar, Rajasthan.

Chapter 6
Prasanta Chandra Mahalanobis (1893–1972)

6.1 Birth, Family, Education

Prasanta Chandra Mahalanobis (P. C. Mahalanobis) was born in the city of Calcutta in a cultured and educated Bengali Brahmo family on 29 June 1893. His father Prabodh Chandra was an artistic person with a special love for painting. But professionally he was associated with the family medicine shop. He gave up his ambitions of becoming a painter and became a successful businessman and gradually took over the responsibility for the big joint family. Prabodh Chandra had a host of friends who were successful in their respective professions and some were well known celebrities of the time. They included important doctors like Kedarnath Das, Narendranath Basu, Upendranath Brahmachari and philanthropists like Subodh Mallick, writers like Satyendranath Dutta, Upendra Kishore Roy Chowdhury and many others. Prasanta Chandra's mother Nirodbashini Devi was the younger sister of the famous physician Dr Nilratan Sircar. However, he lost his mother in 1907, when he was only fourteen years old. Thereafter, his father looked after him and his five other siblings with much care and affection.

Subodh Chandra Mahalanobis, the younger brother of Prabodh Chandra was a great researcher in the field of physiology. He studied physiology and started doing research in that discipline in the Research Laboratory of the Royal College of Physicians at the University of Edinburgh. He graduated from there, and in 1898 he was elected a Fellow of the Royal Society of Edinburgh. That same year, he was appointed Professor and Head of the Department of Physiology at the University of Cardiff at Wales in the UK. In fact, he was the first Indian who held the chair Professorship of Physiology in a British University. After returning to India, in 1900 he set up the department of Physiology at the famous Presidency College in Calcutta. He thus became a pioneer by introducing Physiology as a separate discipline in non-medical colleges of India. Later he also founded the Post-Graduate Department of Physiology and took charge as the Head of the Department in the University of Calcutta. He was a famous scientist, with a large number of research publications to his credit. The

P. Mukherji, *Notable Modern Indian Mathematicians and Statisticians*,
https://doi.org/10.1007/978-981-19-6132-8_6

paternal Uncle Professor Subodh Chandra had a great influence on young Prasanta Chandra.

Prasanta Chandra (P. C. Mahalanobis) studied in the Brahmo Boys' School set up by his own grandfather Guru Charan Mahalanobis. He passed the Entrance examination of the Calcutta University in 1908. The same year he joined the famous Presidency College of Calcutta and in 1912, he graduated from there with Honours in Physics. He got a first division in both the Entrance and the Intermediate examinations. But there was nothing spectacular in his performance. The prescribed syllabi in school or college never really interested P. C. Mahalanobis. During his student days, he could never confine himself to text books or to books related to the subjects of his studies. He had a wide range of interests in subjects as diverse as Astronomy, Architecture, Logic, Philosophy and Psychology. According to his own admission, Prasanta Chandra's father gave him absolute freedom in the matter of education and the son was ever grateful to the father for that.

In March, 1913, even before his Honours results were out, P. C. Mahalanobis sailed to England to pursue higher studies there. Initially he had decided to study in London. But after a brief visit to Cambridge, he decided to study there. Finally he took admission in the King's College, Cambridge and started studying Mathematics. But his result in Tripos (I) examination in Mathematics was very dismal. He barely managed to get a third class. After that he changed his field of study and in 1915 sat for the Tripos (II) in Physics. He did very well and was the only Indian till then to obtain a first class in Natural Sciences in Tripos (II).

After completing his 'Tripos' (part II) Examination he worked at the famous Cavendish Laboratory for some time. He even had plans to work under C. T. R. Wilson and Sir J. J. Thomson. But those plans did not materialize. During his stay in England, he made some important friendships. Rabindranath Tagore had won the Nobel Prize in 1913. P. C. Mahalanobis had the great good fortune of getting close to Tagore, as the poet was spending some time in London then. The young Prasanta Chandra was in the good books of the poet and had free access to him. This friendship between them, in spite of the difference in their age lasted as long as Tagore lived. Tagore played a very important role in all the future activities of P. C. Mahalanobis and the young man too assisted Tagore in all his ventures.

P. C. Mahalanobis during his stay in Cambridge came in contact with the iconic mathematical genius Srinivasa Ramanujan. They became good friends. P. C. Mahalanobis admired Ramanujan throughout his life.

The most important event that happened during the end of his stay in Cambridge was his involvement and interest in Statistics. According to Professor C R Rao, 'At the time of Mahalanobis's departure to India from Cambridge the First World War was on and there was a short delay in his journey. He utilised this time by browsing in the King's College library. One morning, Macauley, the tutor,……..drew his attention to some bound volumes of Biometrika…Mahalanobis got so interested that he bought a complete set of Biometrika volumes….He started reading the volumes on the boat during his journey and continued to study and work out exercises on his own during spare time after arrival in Calcutta'.

Another person who influenced and inspired P. C. Mahalanobis a lot was Professor Brajendranath Seal, Professor of Philosophy at Calcutta University during 1912–1921. He encouraged P. C. Mahalanobis to take up studies in Statistics. Professor Seal as the Chairman of a Committee for Examination Reforms in Calcutta University involved P. C. Mahalanobis in the computational work related to the exercise of 'Examination Reforms'. P. C. Mahalanobis acknowledged his debt to him throughout his life. In a letter dated 2 June 1935, addressed to Prof. Seal, he wrote: 'I may say broadly that I owe to you the entire background of my statistical knowledge, especially in its logical aspects. (I may mention here in passing that also owe a good deal to Mr. C. W. Peake on the technical side, for he introduced me, very early, I think, in 1915, to Karl Pearson's Biometric Tables, Part I, of which had been published just about that time, and the first copy of which had been brought out to India by Mr. Peake).

My first introduction to actual statistical analysis (in its modern mathematical sense) was in connection with your work on the percentage of passes in the Calcutta University ......In the course of this work you made a most comprehensive survey of the whole question, using detailed investigations of the frequency distributions of marks in the different university examinations in different years, correlations between marks, percentage of passes in different years, rate of continuation of higher studies, rate of wastage etc., with separate study for women and I believe students of certain selected communities. I was in close touch with the actual computational work...

In your address before the Races Congress at Rome....you had pointed out very clearly and emphatically the need of using n-dimensional hyper space for the representation of racial types. You had also discussed this idea with me on many occasions ...your idea was the starting point of my own work on the Generalised Distance between Statistical Groups. I have started work on the subject again and I am confident that your idea of a statistical field, when fully developed will lead to results of great importance....'.

6.2 Career, Research Contributions

After returning to India from England in 1915, P. C. Mahalanobis was taken by his famous paternal uncle Professor S. C. Mahalanobis (who was then a Professor and Head of the Physiology Department at Presidency College) to meet the Principal of the Presidency College. Young P. C. Mahalanobis was offered the post of an Assistant Professor against a leave vacancy at the Physics Department. Thereafter he formally joined the Indian Educational Service (IES) and continued working as a regular Lecturer in the Physics Department of Presidency College. It was after 7 long years in 1922, he became a Professor in the Department of Physics. Later he became Principal of Presidency College and retired in 1948. In addition to his duties at Presidency College, Professor Mahalanobis also worked as 'Meteorologist' in charge of the 'Eastern Region', posted in Calcutta from 1922 to 1926.

From 1924, Prof. Mahalanobis single-handedly started working on various meteorological and other problems using statistical methods. With the help of one person S. K. Banerjee, who was recruited temporarily, Prof. Mahalanobis started working on the preparation of the Report on 'Floods in North Bengal'. They had no separate room and very few types of equipment for carrying out large computations. Undaunted by the prevailing difficulties Prof. Mahalanobis, carried on with the work. In 1928, Professor P. C. Mahalanobis was officially requested by the Governments of Bihar and Orissa to prepare a Report of rainfall and floods. The request being official, with approval and permission from the College authorities Prof. Mahalanobis managed to set up a small statistical laboratory in a partitioned corner of a passage of the Baker Laboratory of the Presidency College. He was also able to recruit a few assistants with a very meagre financial remuneration which he paid from his own pocket. That was the first step towards the establishment of Prof Mahalanobis's dream institute.

But before proceeding any further, it is more befitting to discuss Professor Prof. Mahalanobis's main areas of research and contributions. Professor Mahalanobis had received a permanent place for himself in the history of Theoretical and Applied Statistics because of his seminal contributions in the areas of 'Multivariate Analysis', theory of errors in field experiments, and the theory and application of 'Large-scale Sample Survey'.

In India, Statistics was almost unknown in the first quarter of the twentieth century. The subject was not taught in any University, neither was any research carried out. But driven by his personal passion for the subject, Prof. Mahalanobis ventured to work alone in this new area without any help from any quarter. He spent more than half a century of his active life in carrying on research. Broadly, his research career may be divided into three periods.

The first period is 1919–1932. During this time he wrote a total of 38 research papers. Of these 37 were results of the single-handed work carried out by P. C. Mahalanobis. During this period he worked on nine different areas.

In contrast, the second period 1933–1951 reflects research work of a more collaborative nature. The total number of papers published during this period was 198 in as many as 16 different problem areas. This was undoubtedly his most creative period. He had some brilliant young people collaborating with him at that time. The list includes S. S. Bose, R. C. Bose, S. N. Roy, K. R. Nair, anthropologist D. N. Mazumdar and C. R. Rao.

In the last period 1952–1971 the total research output of Prof Mahalanobis with some of his collaborators was 58.

Mahalanobis Distance:

The greatest contribution of Prof Mahalanobis is what is known as D^2-statistic or 'Mahalanobis Distance'.

During the Nagpur Session of the Indian Science Congress in 1920, N. Annadale, the then Director of Zoological Survey of India, requested P. C. Mahalanobis to analyse the anthropometric measurements on the community of people having mixed British and Indian parentage. He handed over the collected data to Prof Mahalanobis. Even earlier Mahalanobis had developed an interest on anthropometric studies by reading articles published in the journal Biometrika. So he lost no time and made extensive and elaborate studies and analysis. In the process he found a way of comparing and grouping populations using a multivariate distance measure. It is denoted as 'D^2' and has been eponymously named 'Mahalanobis distance' by Sir R. A. Fisher, FRS (1890–1962) who is an internationally famous statistician of all times. The research paper titled '*Anthropological observations on Anglo-Indians of Calcutta*' [Rec. Indian Museum, 23, (1922), 1–96] published in 1922 was Mahalanobis's first paper in this discipline. This was a remarkable piece of research work. It established Prof. Mahalanobis's skill in applying statistical methods for extraction of information from given data and make inferences accordingly.

From 1922 to 1936, Professor Mahalanobis wrote and published a total of 15 papers dealing with race-origin, race-mixture and other anthropometry related topics using statistical concepts and tools. These investigations led to the formulation of the statistic known in modern statistical literature as 'Mahalanobis distance' and used widely in problems related to taxonomical classification.

In a paper titled '*On the need for standaradization in measurements on the living*' [Biometrika, 20 A, (1928), 1–31], Mahalanobis examined anthropometric data collected from different sources and showed that because of differences in the definitions of measurements in different investigations, a comparative study could not be made.

His celebrated paper published in 1936 and titled '*On the generalised distance in statistics*' [Proceedings of National Institute of Sciences. 2, (1936), 45–55] is perhaps his best paper in the field of D^2-statistic. In the same year, he in collaboration with S. N. Roy and R. C. Bose published another important paper in Sankhya, titled '*Normalisation of variates and the use of rectangular coordinates in the theory of sampling distributions*' [Sankhya, 3, (1936), 1–40]. P. C. Mahalanobis independently published a paper titled '*Normalisation of Statistical variates and the use of rectangular coordinates in the theory of sampling distributions*' [Sankhya, 3, (1937), 35–40].

Prof. Mahalanobis's early work on statistical analysis of anthropometric data raised many theoretical problems, which in turn opened up wide field for research in 'Multivariate Analysis'. His three students R. C. Bose, S. N. Roy and C. R. Rao made outstanding research and notable contributions in this area.

In 1945, Professor Mahalanobis in association with D. N. Majumdar, who was an anthropologist and C R Rao wrote a very important research paper titled '*Anthropometric Survey of United Providences 1941; A statistical study*' [Sankhya, 9, (1949), 89–324]. This research paper is of great scientific excellence.

Field Trials:

Another area of research by P. C. Mahalanobis deserves special mention. He wrote singly or in collaboration with others 50 research papers and reports on topics related to agronomy. A sizable number of them were of the nature of field trials. Since the 1920s P. C. Mahalanobis was carrying out such field experiments all by himself with practically no help from anyone.

World renowned statistician R. A. Fisher of UK was doing almost the same kind of work in Rothamstead station. However, when P. C. Mahalanobis started his field experiments, he was quite unaware of Fisher's work. It was R. A. Fisher, who after seeing some of Mahalanobis's publications in this area, took the initiative and sent him some of his own papers to the latter. Thus a professional contact was established between them. It was a long-lasting relationship and became very helpful to Professor Mahalanobis in many ways.

Statistical Tables:

In the field of 'Theoretical Statistics' Prof. P. C. Mahalanobis made major contributions by computing and tabulating them in the form of 'Statistical Tables'. In 1932, he was the first person who tabulated the derived results of R. A. Fisher related to the one percent and the five percent points of the distribution. A comment of statistician K. R. Nair is important from the historical context. He wrote:

> *His (P. C. Mahalanobis) first tabulation of five and one percent points appeared in 1932. Today this quantity is commonly referred to as Snedecor's F, since Snedecor, in ignorance of the Indian work later tabulated the variance ratio under the symbol F, chosen in honour of Fisher. In Statistical Medical Research (1938), Fisher avoided using F, since this symbol was not used in the tabulation of Mahalanobis, which had priority.*

Professor Mahalanobis constructed 3 more tables in 1933. Out of which *'Tables for L-tests'* was published in Sankhya (Volume-I). The *'Tables for Random Samples from a Normal Population'* was published in Sankhya (Volume-I) in 1934.

Large-scale Sample Survey:

Another area where Professor P. C. Mahalanobis made seminal contribution is in 'Applied Statistics' by developing the 'Theory of Sample Survey'.

The first recorded reference to Professor Mahalanobis's involvement with this discipline goes back to 1932, when he delivered a talk titled 'Elements of the theory of Sampling' at the Statistical Laboratory of the Presidency College, Calcutta.

He defined 'Sampling' as follows:

'A sample consists of two or more elementary units drawn from a population (universe or field) in a random manner through the sample frame and the inference about the population is based on the observations, measurements and experiments on the variates (or characteristics) of the elementary units in the sample.'

Right from the beginning, Professor Mahalanobis had clearly realized the immense potentialities of the method of random sampling and also the necessity for correct interpretation of the results emanating from sample surveys.

Mahalanobis started using the method of sample survey in his scheme for sample census of jute crop in Bengal way back in 1937. The Internationally famous Statistician Professor Harold Hotelling (1895–1973) of USA in 1938 after carefully examining Mahalanobis's scheme for a sample census in Bengal wrote: '..........no technique of random samples has, so far as I can find, been developed in the United States or elsewhere, can compare in accuracy or in economy with that described by Professor Mahalanobis.' Professor Mahalanobis proceeded to popularize large-scale sample surveys by actually demonstrating its usefulness. Reviewing Professor Mahalanobis's outstanding work in the field of Sample Surveys, Dr Y. P. Seng in 1951, wrote: 'India can therefore safely claim to rank with the United States as among the foremost users of the sampling method in social and economic research. And it is a very happy combination, for in the United States we have a typical example of an industrial and highly developed country while in India the conditions approximate more nearly to those of a country not so highly developed or more specifically to the conditions of those countries, which, like China, have no genuine statistics and where such statistics, if they are to be obtained at all, have to be obtained mainly by sample surveys, for which the experience of India will serve as a guide and as an example worthy of imitating.'

Professor F. Yales, FRS (1902–1994) in 1951, remarked, in the context related to the work of the United Nation's Sub-Commission on Statistical Sampling: 'He (Mahalanobis) consequently recognised more clearly than most, that if more world censuses were to be properly carried out in the less developed countries, the use of sampling method would be essential and it was he who proposed the setting up of the Sub-Commission on Statistical Sampling in order to assist the work of proper application of Sampling methods.'

And finally the remarks of Professor Sir R. A. Fisher, FRS, in 1962, during the first convocation of the Indian Statistical Institute: 'I need hardly say that I refer to the emergence of a statistically competent technique of sample survey, with which I believe Professor Mahalanobis's name will always be associated. What at first most strongly attracted my admiration was that the Professor's work was not imitative'.

Random sampling and sample survey techniques are the bedrocks of applied statistics. Professor Mahalanobis himself and his students such as D. B. Lahiri, J. M. Sengupta, M. N. Murthy, S. K. Banerji and others did commendable work using those methods in various areas such as crop survey, agriculture, demography, rainfall and flood related studies, fisheries, industry and many other socio-economic fields.

It should be noted that 'Large-scale Sample Survey' techniques which are used today are largely based on the pioneering work done by Prof. P. C. Mahalanobis and his team in the forties and fifties of the twentieth century. This methodology for conducting large-scale sample survey was developed during 1937–1944 in connection with the numerous surveys which were planned and executed under the leadership of Mahalanobis by the 'Indian Statistical Institute'. The basic results on large-scale sample survey were published by Prof. P. C. Mahalanobis in his paper titled '*On large scale sample surveys*' [Phil. Transactions of the Royal Society, London, Series B, (1944), 329–451]. It was also presented by him at a 'Meeting' of the 'Royal Society', London. His three major contributions to sample survey techniques are: (i) pilot surveys, (ii) concept of optimum survey design and (iii) interpenetrating network of samples.

It may be noted here that the excellence of Professor Mahalanobis's scientific output lies in the inseparable relation that is represented between theory and its application. He believed that 'Statistics' was a key technology that could be used for problems arising in different Scientific, Sociological and Economic situations. All through his life in all research contributions, he remained firm in this conviction.

Operations Research:

Another area where Prof. Mahalanobis worked in the early years of his career relates to river floods. This work is in the nature of 'Operations Research' (OR). This became a separate discipline after World War II. He worked on problems of floods on various Indian rivers such as Damodar, Brahmini, Mahanadi, etc. In fact, he was the first man who made the first calculations and gave the idea for a multipurpose (flood control, irrigation and power) scheme for the 'Mahanadi river system' in Orissa (present-day Odissa). This formed the basis of the 'Hirakud Hydroelectric Project' which started functioning in 1957.

Fractile Graphical Analysis:

During the last decade of his life Prof. Mahalanobis developed a semi-nonparametric method of comparison of two samples. He named it 'Fractile Graphical Analysis'. This technique has widespread applications in socio-economics, demography, psychology, biometry, etc.

Errors in field records and their corrections:

Before ending the discussion on Professor P. C. Mahalanobis's contribution in various branches of statistics, it is necessary to mention one particular trait in his working career. Prof. Mahalanobis laid great emphasis on the need for thorough scrutiny of field records and derived systematic methods for detection of recording error and for making necessary adjustments. He was extremely meticulous about the correctness and dependability of collected data. He was extremely careful about errors. He had a special type of interest in some theoretical problems such as: errors of observation, errors of measurements, sampling errors, etc. If ever he had any doubts about the collected data or related calculations, he would check the whole problem on his own, step by step and single-handedly. He was also very particular about the way a data should be presented. Later on, when the students came to know about Professor Mahalanobis's obsession with error-free observations and measurements, etc., they constructed a sobriquet with his initial PCM. They privately referred to him as 'Professor of Counting and Measurements'.

Setting up Institutions:

Prof. P. C. Mahalanobis was a giant among the institution builders of India. His greatest achievement was establishing the Indian Statistical Institute (ISI) in Calcutta. He founded the Institute in 1931 with the aim of providing facilities for high level research, teaching, training and execution of large-scale project work. A brief historical account about the establishment of ISI and the opening of a separate Department of Statistics for Post-Graduate courses in the Calcutta University has been given in the chapter on 'Historical Prelude'. His role in setting up the 'National Sample Survey' (NSS) is undoubtedly one of his outstanding achievements.

Prof. P. C. Mahalanobis was the 'Chief Executive Secretary' of the Indian Statistical Institute, Calcutta from the time of its foundation till the end of his own life. Under his dynamic leadership, ISI was able to successfully establish its credibility as a world-class institution, promoting research and training at the interface of Statistics and other disciplines, which comprised physical, biological, geological and socio-economic sciences. He had the unique capability of identifying talented people and thus recruited the brightest minds to join and work at ISI. As already mentioned earlier, in 1959, the Indian Parliament declared ISI an 'Institute of National Importance'. The ISI, Calcutta then introduced undergraduate, graduate and research programmes and also started awarding B. Stat, M. Stat and Ph.D. degrees in Statistics. Prof. Mahalanobis was also aware of the importance of international collaboration. He invited eminent researchers from other countries to visit the Institute. As the fame of ISI spread internationally as an excellent centre of research in Statistics, famous scientists and statisticians from all over the world came and spent time at ISI, Calcutta. Some of the eminent personalities include J. L. Doob. R. A. Fisher, Ragnar Frisch, J. K. Galbraith, J. B. S. Haldane, A. N. Kolmogorov, S. Kuznets, R. Stone, A. Wald and N. Wiener.

In the newly set up ISI, under the bold leadership of Prof. P. C. Mahalanobis, the new techniques of random sampling was used for estimating acreage and yield of jute crops in Bengal in 1937. Again it was Prof. Mahalanobis, who was the guiding spirit for organizing the first conference on Statistics, the 'Indian Statistical Conference' in 1938. He used all his persuasive powers and patience to convince the authorities of Calcutta University for the need of setting up a separate Department in Statistics and it actually started functioning from 1941. Prof. P. C. Mahalanobis was the Honorary Head of the Department. Under his influence, for the first time in the history of Indian Science Congress, Statistics was included as a separate discipline in the 'Mathematics Section' in 1942. Subsequently a separate section for Statistics was started from 1945. After India attained her independence, the first Prime Minister of India Jawaharlal Nehru realized the importance of using statistical techniques for better economic planning and implementation. So in 1949, a 'Central Statistical Unit' was initiated by the Government of India with Prof. P. C. Mahalanobis as the 'Honorary Statistical Adviser' to the cabinet. For greater administrative efficiency, in 1951 the 'Central Statistical Organization' (CSO) was created by the central government. The function of the newly created agency was to coordinate all statistical activities of the government. Shortly after that a separate Department of Statistics was formed. Jawaharlal Nehru had great faith on and was very close to Prof. Mahalanobis. So as advised by Prof. Mahalanobis, the National Sample Survey (NSS) was established in 1950. Professor P. C. Mahalanobis was a key figure in setting up the National Sampling Survey (NSS) and the Central Statistical Organisation (CSO). With the experience gained from Bengal earlier, he wanted to implement it throughout the country. One may unhesitatingly say that one of the greatest achievements of Prof. P. C. Mahalanobis was the establishment of the NSS in 1950. According to Prof. C. R. Rao:

> It is a continuing exercise employing a large number of field investigators spread over the entire country for collection of information on socio-economic and demographic aspects of the population periodically on a sampling basis............The task proved to be extremely difficult especially because there was no parallel anywhere in the world in the use of continuing sample survey covering an entire country for collecting official statistics. Mahalanobis was

concerned more with the practical aspects of sample surveys rather than with mathematical research on sample survey. As Chairman of the United Nations Sub-Commission on Sampling (1947–51), he advocated the use of sample survey methods in less developed countries for the collection of socio-economic and demographic data and laid down specifications for conducting large-scale sample surveys.

In 1949, Prof. Mahalanobis was made the Chairman of the 'Indian National Income Committee'. This prompted him to think about the macro-economic problems of India. He discovered that there were gaps in information in the computed national income. Actually to seal these gaps, he proposed the formation of NSS in 1950. He also formed a statistical unit devoted to the study of national income. The idea of 'Quality Control' was initiated in India by Prof. P. C. Mahalanobis for improving the quality of industrial products and standardization of the price structure of finished items.

Apart from all these important responsibilities, at the request of the then Prime Minister Mr. Nehru, Prof. Mahalanobis became associated with the 'Planning Commission'. He was deeply involved in the formulation of India's 'Second Five Year Plan'. In all these activities, he was a close adviser of the Prime Minister. While handling these monumental responsibilities, Prof. Mahalanobis became aware of wider national and international problems. During that time he wrote extensively on subjects like:

- 'Priority of basic industries'
- 'Role of scientific research, technical manpower and education in economic development'
- 'Industrialization of poorer countries and world peace'
- 'Labour problems, unemployment and demographic problems'.

6.3 Special Honours and Awards

Prof. P. C. Mahalanobis was an internationally acclaimed statistician and the 'Father of Statistics' in India. He was elected as 'Honourable Fellows' to numerous famous societies and received many awards and accolades from home and abroad.

He was the Founder Fellow of the 'National Academy of Sciences' [presently known as 'Indian National Science Academy' (INSA)]. Later he also served as the President of INSA during 1957–1958. He was also an elected 'Fellow' of the 'Indian Academy of Science', Bangalore. In 1945, he was elected 'Fellow' of the 'Royal Society, London' (FRS). He was elected 'President' of the annual 'Indian Science Congress' held at Pune in 1950. In 1951, he was elected 'Fellow' of the 'Econometric Society, USA'. He was elected 'Fellow' of the 'Pakistan Statistical Association' in 1952. He was elected 'Honorary Fellow' of 'Royal Statistical Society, UK' in 1954. In 1957 he became the 'Honorary President' of the 'International Statistical Institute'. He was elected as a 'Foreign Member' of the 'USSR Academy of Sciences' in 1958. He was elected 'Fellow' of the King's College, Cambridge University in 1959. He was elected a 'Fellow' of the 'American Statistical Association' in 1961.

In 1963 Prof. Mahalanobis was elected a 'Fellow Member' of the 'World Academy of Arts and Science.' In 1968, the Government of India awarded him the prestigious 'Padmabibhushan' award for his contributions to Science and services to his own country.

Prof. P. C. Mahalanobis was the recipient of several 'Medals' for his outstanding contributions to science and society. The 'Weldon Medal' from Oxford University was awarded to him in 1944. In India, he received the 'Sir Deviprasad Sarvadhikari Gold Medal' in 1957, the 'Durgaprasad Khaitan Gold Medal' in 1961 and the 'Sriivasa Ramanujan Gold Medal' in 1968 for his contributions in Science. Prof. P. C. Mahalanobis was honoured with the 'Czechoslovak Academy of Sciences Gold Medal' on the occasion of his 70th birthday in 1963. He also received many 'Honorary Doctorate' degrees from various Universities; including the University of Calcutta and the Delhi University'.

Prof. P. C. Mahalanobis passed away in Calcutta a day before his 89th birthday, on 28 June 1972, due to some post-operative complications. In his death Bengal and India lost one of their greatest statisticians of all times. In his 'Obituary', another doyen of Statistics and a close associate of Prof. Mahalanobis, Prof. C. R. Rao wrote:

> Statistical science was a virgin field and practically unknown in India before the twenties. Developing statistics was like exploring a new territory. It needed a pioneer like Mahalanobis, with his indomitable courage and tenacity to fight all opposition, clear all obstacles and throw open wide pastures of new knowledge for advancement of science and society. With the passing away of Prof. Mahalanobis, India has lost an outstanding personality, the like of whom is born perhaps once in several generations.

6.4 Important Milestones in the Life of Professor Prasanta Chandra Mahalanobis

1893: Born in Calcutta on 29 June 1893.

1908: Passed the Matriculation Examination from Brahmo Boys' School, Calcutta.

1912: Passed Physics (Honours) with a First class from Presidency College, Calcutta.

1913: Proceeded to England for pursuing higher studies.

1913: Joined the King's College, University of Cambridge, England.

1915: Returned to India after completing Tripos (Part II) Examination securing a First class.

1915: Joined the Presidency College, Calcutta first in a leave vacancy. Then he joined the 'Indian Educational Service' (IES), as a permanent Faculty Member of the Physics Department of Presidency College, Calcutta.

1922: Became a Professor in the Department of Physics, Presidency College, Calcutta. Discovered the famous D^2-statistic, known in modern statistical literature as 'Mahalanobis Distance'.

1928: He set up a small 'Statistical Laboratory' in an anteroom of the Baker Laboratory of Presidency College, Calcutta.

1931: On 14 December 1931, a firm decision was taken by Prof. Mahalanobis and some of his academician friends on the setting up of the 'Indian Statistical Institute'. Thereafter the 'Institute' started functioning unofficially under the leadership of Prof. Mahalanobis.

1931–1972: Served as the Chief Executive Secretary, Indian Statistical Institute.

1932: On 28 April 1932, the 'Indian Statistical Institute' (ISI) was officially registered.

1933: Prof. Mahalanobis started publishing the first Journal on Statistics named 'Sankhya' in India. He was also the 'Editor' of the Journal since its inception.

1935: Founder Fellow of the 'National Academy of Sciences' [presently called 'Indian National Science Academy' (INSA)]. The same year he was also elected 'Fellow' of the 'Indian Academy of Sciences', Bangalore.

1942: The British Government bestowed 'Order of the British Empire' (OBE) on Prof. Mahalanobis. The University of Oxford awarded the prestigious 'Weldon Memorial Prize' and Gold Medal.

1945: Elected 'Fellow' of the 'Royal Society of London' (FRS).

1947–1951: Chairman of the 'United Nations Sub-Commission on Sampling'.

1949: 'Honorary Statistical Adviser' to the Government of India.

1950: Establishment of the 'National Sample Survey' (NSS).

1950: Elected 'President' of the annual 'Indian Science Congress' held at Pune.

1951: Elected 'Fellow' of the 'Econometric Society, USA'.

1952: Elected 'Fellow' of the 'Pakistan Statistical Association'.

1954: Elected 'Honorary Fellow' of the 'Royal Statistical Society, UK.'

1957: Elected 'Honorary President' of the 'International Statistical Institute'.

1957–1958: President of the 'Indian National Science Academy'.

1958: Elected 'Foreign Member' of the 'USSR Academy of Sciences'.

1959: Elected 'Fellow' of King's College, Cambridge, England.

1961: Elected 'Fellow' of the 'American Statistical Association'.

1963: Elected a 'Fellow Member' of the 'World Academy of Arts and Science.'

1963: Was awarded the 'Czechoslovak Academy of Sciences Gold Medal', on the occasion of his 70th birthday.

1968: The Government of India awarded him the prestigious 'Padmabibhushan' award for his contributions to Science and services to India.

1968: Awarded the 'Srinivasa Ramanujan Gold Medal' for his contributions to Science.

1972: Passed away on 28 June 1972 in Calcutta due to some post-operative complications.

Chapter 7
Nikhil Ranjan Sen (1894–1963)

7.1 Birth, Early Education

Nikhil Ranjan Sen was born on 23 May 1894. His father was Sri Kalimohon Sen, and mother was Bidhumukhi Devi. He was born in the district of Dacca, now in Bangladesh. His school education started in Dacca Collegiate School, but he passed his Entrance Examination from Rajsahi Collegiate School in 1909. Later he studied in Presidency College, Calcutta. Meghnad Saha and Satyendranath Bose were both his classmates right from the year 1911. In Presidency College, N R Sen had the privilege of being taught by such illustrious Professors as Acharya Jagadish Chandra Bose and Acharya Prafulla Chandra Roy. Undoubtedly, the young talented students like M. N. Saha, S. N. Bose and N. R. Sen were greatly influenced by these great scientists. Both the Acharyas very successfully inculcated the spirit of enquiry into these young minds. In 1913, after obtaining a first class in B.Sc. (Mathematics Honours), N. R. Sen joined M.Sc. class in the Department of Applied Mathematics (then called Mixed Mathematics) of the Calcutta University. The classes however used to be held in the Presidency College. N. R. Sen took his M.Sc. degree in 1916 securing First class first position.

7.2 Career, Higher Education, Research Contributions

In 1917, N. R. Sen was appointed for a teaching position in the newly set up Department of Applied Mathematics of the University of Calcutta by the illustrious mathematician Vice-chancellor Sir Asutosh Mookerjee. Sir Asutosh, as was his style, inspired and encouraged Nikhil Ranjan Sen to take up serious research work in the field of Mathematics. In the initial years of his career, N. R. Sen took a keen interest in the theory of Newtonian Potential and also in the mathematical theories of Elasticity and Hydrodynamic waves. He published a number of research papers

© The Author(s), under exclusive license to Springer Nature Singapore Pte Ltd. 2022
P. Mukherji, *Notable Modern Indian Mathematicians and Statisticians*,
https://doi.org/10.1007/978-981-19-6132-8_7

in these areas and published them in well-known national and international journals. In 1921, he obtained the D.Sc. degree from the University of Calcutta for his original contributions in Applied Mathematics. Almost immediately after this, the Calcutta University granted him study leave and he proceeded to Germany to pursue further research. There he worked at the newly set up 'Institute for Physics' in Berlin with Professor Max Von Laue (1879–1960). Under Professor Von Laue, N. R. Sen carried out research in the General Theory of Relativity and on Cosmological problems. During his stay in Europe, Nikhil Ranjan worked at various famous centres of Advanced Study and research in Mathematics in the continent, namely Berlin, Munich and Paris. During these visits, N. R. Sen came into contact with great scientists like Max Planck, Arnold Sommerfeld, Albert Einstein and Louis de Broglie. He utilized these opportunities and seriously took up studies in the newly developed subject of Quantum Mechanics. He also learnt the Mathematical Theory of Probability and the theory of Topology during his stay in Germany. A remarkable fact is that though Nikhil Ranjan Sen had been trained as an Applied Mathematician, he had great interest in various branches of Pure Mathematics too.

His research activities were also spread over a wide range of subjects and topics. He did some notable work on some problems of Spherical Harmonics. Professor Ganesh Prasad, the first Rashbehary Ghosh Professor of Applied Mathematics influenced N. R. Sen in this respect. In the early thirties, Professor Sen did considerable amount of work on wave mechanics which included Dirac's relativistic equations. He took up problems in Cosmology. He investigated the relativistic effects in stellar bodies. Around 1940, after Bethe's law of energy generation was established, N. R. Sen started working on problems of stellar constitution. In India he did pioneering work on the internal constitution of stars. In collaboration with his students he constructed stellar models based on the theoretical laws of thermo-nuclear energy generation.

From 1950 onwards till the end of his life, N R Sen did a lot of original research in the field of Fluid Dynamics. In India, in the area of Fluid Dynamics he did pioneering research work on Boundary Layer, Wave Resistance, Isotropic Turbulence and Shock Waves. His investigations on Heisenberg's equation for the decay of isotropic turbulence deserve special mention.

He built up a strong team of research workers notable among his students being U. R. Barman, N. L. Ghosh and T. C. Roy. Personally Prof. N. R. Sen published 46 research papers in various well-known national and international journals.

After his return from Europe, in 1924, Dr. N. R. Sen took over as the 'Rashbehary Ghosh Professor' of Applied Mathematics in the University of Calcutta. That was the beginning of a glorious era for the Department of Applied Mathematics. Professor N. R. Sen introduced new subjects in the Post-Graduate curriculum and inspired his young colleagues and research associates to take up original and challenging problems and solve them in modern areas like Relativity, Astro-physics, Quantum Mechanics, Geophysics, Statistical Mechanics, Fluid Dynamics, Magneto Hydrodynamics, Elasticity and Ballistics. He was the fountain-head of inspiration to research workers in Calcutta University, and under his dynamic leadership the Department of Applied Mathematics became a vibrant centre of research and teaching in Applied

Mathematics and earned a reputation throughout the country which would be hard to match. For obvious reasons, he was called the 'Father of Applied Mathematics in India'.

Prof. N. R. Sen's research work involved theoretical modelling as solutions of Einstein's equations with physical properties. Some of the well-known ones are:

I.   Static solution with spherical symmetry in most general form,
II. Spherical shell with surface matter,
III. Relationship to de Sitter Universe: A transformation of the solution inside the shell leads to the de Sitter model,
IV. Equilibrium of a charged particle having a definite spherical boundary.

He was instrumental in setting up a laboratory for numerical calculations in the Department of Applied Mathematics. He was also responsible for setting up the hydrodynamical laboratory in the Department with a flume and a wind tunnel. He was well aware of the necessities for conducting research on par with advanced countries. In spite of having access to limited resources, he tried his utmost to make the department as well equipped as possible.

In spite of his very hectic schedule which involved teaching, research and administration of the Department of Applied Mathematics, Prof. N. R. Sen served on various committees and advisory bodies of scientific and educational academies and agencies all over India.

He was the founder fellow of the National Institute of Sciences ever since its inception in 1935. Throughout his life he was deeply engaged with the Indian Association for the Cultivation of Science in Calcutta and Calcutta Mathematical Society. He worked very hard to raise the standard of the 'Bulletin' published by the Calcutta Mathematical Society. He was also associated with the establishment of the 'Indian Statistical Institute' in Calcutta right from the beginning.

Prof Sen made remarkable contribution in various disciplines of Applied Mathematics. His complete list of publication amply justifies this statement. But apart from his personal contributions in scientific research, his role in modernizing the Department of Applied Mathematics, inspiring scores of students to take up research in different area of Applied Mathematics and enhancing the prestige of the Department as the model centre of excellent scientific research will be written in the annals of Calcutta University in golden letters. His memory is cherished by any serious research worker of Applied Mathematics not only in Bengal but all over the country.

7.3   Important Milestones in the Life of Professor Nikhil Ranjan Sen

1894: Born in the district of Dacca (in erstwhile East Bengal, present-day Bangladesh) on 23 May 1894.
1909: Passed the Entrance Examination from Rajshahi Collegiate School, in erstwhile East Bengal.

1913: Passed B.Sc. Mathematics (Honours) with a First class from Presidency College, Calcutta.

1916: Passed M.Sc. in Mixed Mathematics from the University of Calcutta. He stood First with a First class.

1917: Appointed as a Lecturer in the newly set up Department of Mixed Mathematics (later named Applied Mathematics), Calcutta University.

1921: Obtained the D.Sc. degree from the Calcutta University for his original contributions in Applied Mathematics.

1921: Left for Germany to pursue further research. There he carried out research in collaboration with Professor Max Von Laue.

1924: Returned to Calcutta and joined as 'Rashbehary Ghosh Professor' of Applied Mathematics in the University of Calcutta.

1935: Founder Fellow of the 'National Institute of Sciences' [Later known as Indian National Science Academy (INSA)].

1963: Died in Calcutta on 13 January 1963, after a prolonged illness.

Chapter 8
Suddhodan Ghosh (1896–1976)

8.1 Birth, Family, Education

Suddhodan Ghosh was born on the 26 July 1896, in a well-to-do and aristocratic Bengali family in Calcutta. His father Sarat Chandra Ghosh came from a rich family and was himself an affluent landlord. His mother Patitpabani Debi was a kind-hearted, generous and devout lady. His entire childhood has been described well by a remark made by S. Ghosh himself. He said: 'I was born a weakling. My delicate constitution did not permit me to participate in any healthy pastime. I could hardly run a few yards for want of breath. The house physician prophesied that I would not outlive a dozen years. Thus I became my mother's pet and pampered child. Thanks to my mother's prayers and penance and through God's grace, I outwitted the doctors and developed into a normal child. Books became my delight, my one absorbing passion. With a passion for books, I developed a love of solitude. Alone and aloof from the excitements of the world, I delved into books and scriptures, watching things profane and things divine from a distance.'

The famous scientist Acharya P. C. Ray was related to S. Ghosh. His grandfather Haranchandra Ghosh was the first Bengali to become the Chief Judge of the Calcutta Small Cases Court. His family and noted relatives had a deep and abiding influence on young S. Ghosh.

Right from his childhood S. Ghosh showed extraordinary aptitude for Mathematics. After completing his early schooling in Metropolitan Institution (established by Ishwar Chandra Vidyasagar), in 1910, S. Ghosh was admitted to the famous Hindu School in class seven. Though his famous grandfather Haranchandra Ghosh was one of the founder-donor of Hindu school, still S. Ghosh was admitted strictly on merit as he had refused admission in 'donor's quota'. In all school examination S. Ghosh always topped the list of the successful candidates. The renowned teacher of Mathematics, U. L. Bakshi had a special liking for S. Ghosh. In 1914, S. Ghosh passed the Matriculation with a first division and secured the 10th position in order of merit.

Soon after this, the Ghosh family met with a number of tragedies. The eldest paternal uncle of S. Ghosh, P. C. Ghosh passed away suddenly. His youngest brother

P. Mukherji, *Notable Modern Indian Mathematicians and Statisticians*,
https://doi.org/10.1007/978-981-19-6132-8_8

converted to Christianity and was forced to leave home. The great 'Ghosh joint family' disintegrated. Meanwhile S. Ghosh joined the Scottish Church College and enrolled for the Intermediate Class in Science. In 1916 he passed the I.Sc [Intermediate Examination in Science] and secured the eighth place in order of merit in the combined list [I.A. and I.Sc.] of successful candidates. He got the highest marks in Mathematics among all students. He joined the famous Presidency College of Calcutta to study B.Sc. in Mathematics (Honours). At this time his parents were facing immense financial problems. One of S. Ghosh's aunt who was married into a rich family helped him financially, so that he could buy the necessary books and a wooden almirah for storing the books, writing materials, notebooks, etc. In the Presidency College, S. Ghosh got the best exposure a student could dream of. From teachers such as Acharya J. C. Bose, Acharya P. C. Ray, Dr. D. N. Mallik, S. Ghosh received the necessary inspiration and guidance that helped him to blossom into a scholar of the highest class. In 1918, S. Ghosh stood First with a First class in Mathematics (Honours) from Presidency College. He was the recipient of the 'University Gold Medal' for his remarkable academic performance. After reaching home, S. Ghosh placed the Gold Medal at his mother's feet. She in turn blessed him with gold mohar of the Akbar era. S. Ghosh cherished his mother's gift till the last day of his life. As his last wish, after his demise the gold mohar was gifted to the University of Calcutta.

In 1918, he joined the Department of Mixed (Applied) Mathematics at the University College of Science, University of Calcutta to continue his Post-Graduate studies. At that time Prof. S. N. Bose was a Faculty Member in the Department and taught the 'Mathematical Theory of Elasticity'. S. Ghosh was so impressed by the way S. N. Bose explained the subject that he developed a life-long fascination for continuum Mechanics. In 1920, S. Ghosh passed the M. Sc. Examination again standing First with a First class in Mixed Mathematics. He secured about 90% of the total marks. He again earned the 'University Gold Medal' for his brilliant scholastic performance. His mother was no more, and she could not see his second Gold Medal.

8.2   Career, Research Contributions

Immediately after completing his M.Sc., S. Ghosh started working as a 'Research Scholar' under the guidance of Prof. N. R. Sen, the then 'Rashbehary Ghosh Professor' and Head of the Department of Applied Mathematics. It may be noted that in those days, the scholarships fetched only Rs. 75/- per month. S. Ghosh had declined an enviable offer from the Imperial Bank of India [Comparable to the present day State Bank of India] and also refused to go abroad for higher studies as suggested by Acharya P. C. Ray and was quite happy with the paltry remuneration he received as a 'Rashbehary Ghosh Scholar'. It was his dream to stay in his Motherland and do research in Mathematics with just sufficient money to meet the basic needs. He personally believed that a life of affluence and luxury was a deterrent to scholarly and intellectual ardour. In 1925 he competed and won the prestigious 'Premchand Roychand Studentship' from the University of Calcutta. He was also awarded 'Sir Asutosh Mookerjee Gold Medal' by the University for his outstanding research contributions. In 1928, S. Ghosh obtained the D.Sc. degree from the University of Calcutta.

At the persuasion of his favourite teacher Prof. S. N. Bose, in 1928 Dr. S. Ghosh went to the Dacca University in erstwhile East Bengal and joined the Department of Mathematics as a Lecturer. But after a brief stay he returned to Calcutta. Soon after that he was appointed as a Lecturer in his alma mater, the Department of Mixed Mathematics, University of Calcutta. He was much admired by his students and colleagues for his deep knowledge in Mathematics and his admirable teaching techniques.

Dr. S. Ghosh was a man of principles; honesty and integrity were synonymous with his character. There are plenty of examples to prove the point. In 1953, Dr. Ghosh was selected for the post of a Reader in the Department of Applied Mathematics (earlier known as Mixed Mathematics). In fact that was the first such post created for the Department. However Dr. Ghosh declined the offer, as he felt that many equally qualified senior colleagues were being bypassed. The University authorities had to ultimately transfer that post to a sister department. Another incident also reflects his belief in truth and honesty. In 1954, Prof. J. C. Ghosh was the Director of the newly established Indian Institute of Technology (IIT) at Kharagpur, West Bengal. He was searching for a brilliant mathematician to become the Head of the Department of Mathematics. Prof. S. N. Bose had strongly recommended the name of Dr. S. Ghosh for the post. In the meantime a contemporary of Prof. S. Ghosh had talked to him and indirectly requested him not to apply, because then he had no chance of being selected. Dr. S. Ghosh had assured his friend that he would not apply. Later when approached by the authorities of IIT, Kharagpur even through his former supervisor Prof. N. R. Sen, he did not apply for the post. He did not divulge the fact that he was keeping himself away because he had made a promise to his friend. But a man wedded to truth and integrity could not be dislodged from his self-chosen path of righteousness.

Dr. S. Ghosh followed the same philosophy throughout his life. If he felt that something was not fair, he boldly opposed it and never cared about the outcome. Another such event occurred in 1958. He had differences of opinion with the then Head of the Department of Applied Mathematics regarding the curriculum and distribution of teaching load among his colleagues in the Department. He promptly submitted his letter of resignation to the University authorities. They however did not accept the letter. They knew his worth and the then treasurer of Calcutta University, Dr. Satish Chandra Ghosh somehow was successful in persuading him to join the Department of Pure Mathematics of the University. The event actually became a boon for that Department. The University authorities realizing the indispensability of Prof. S. Ghosh, soon made him the Head of the Department of Pure Mathematics. What was a great loss for the Department of Applied Mathematics became a great gain for the Department of Pure Mathematics.

After Prof. N. R. Sen retired, the 'Sir Rasbehary Ghosh Chair Professorship' of the Department of Applied Mathematics was vacant and the post was duly advertised.

Dr. S. Ghosh did not even apply. But the post was offered to him by the concerned authority. Then he accepted the offer with his customary modesty. Prof. S. Ghosh acted as the 'Sir Rashbehary Ghosh Professor' and Head of the Department of Mathematics from 1959 to 1963. He did not seek any extension and retired exactly when it was due. During his tenure, he helped and inspired the students in every possible way. He initiated them into challenging unsolved problems and confidently guided them so that they could make original contributions in their own research work. His personal deep knowledge and vision enabled to kindle their creative ability.

Prof. S. Ghosh was famous for his punctuality. In his thirty-five year long teaching career in the University of Calcutta he had availed only 13 days of leave.

As already mentioned, he was a great admirer of Prof. S. N. Bose and was particularly impressed by his teaching. Being fascinated by 'Mechanics of Continuum', Prof. S. Ghosh made valuable contributions in both Fluid Mechanics as well as Solid Mechanics. His additional advantage was his enviable knowledge in 'Pure Mathematics'. His remarkable originality lay in reducing complex engineering problems to those of Pure Mathematics. In fact the first research paper that Prof. S. Ghosh published in 1924 was in the area of 'Fluid Mechanics'. In the paper titled '*On liquid motion inside certain rotating circular arcs*' [Bull. Cal. Math. Soc., 15, (1924), 27–46], his discussion involved solving Laplace's equation in ψ, the stream function and its value on the specified boundaries. Another important paper in this area is titled '*On the steady motion of viscous liquid due to translation of a tore parallel to its axis*' [Bull. Cal. Math. Soc., 18, (1927), 185]. Prof. S. Ghosh solved this problem by using curvilinear co-ordinates (α, β) and assuming a certain type of Stoke's function. In 1928, he published the paper titled '*The steady rotational motion of a liquid within fixed boundaries*' [Bull. Cal. Math. Soc., 19, (1928), 59–66]. Thereafter, Prof. S. Ghosh shifted his attention to problems in the 'Theory of Elasticity'.

His work in Elasticity comprised (i) Bending of Elastic Plates, (ii) Problems of Dislocation, (iii) Plane Problems, (iv) Vibrations of Rings and (v) Torsion and Flexure Problems. He published more than 20 research papers on the topics mentioned above. A brief discussion is being made on his important research contributions.

The problem of bending of a loaded circular plate, the load being uniformly distributed was first solved by the French mathematician S. D. Poisson (1781–1840). Another famous French expert on Mechanics and Mathematics A. J. C. B. De Saint–Venant (1797–1886) studied the case where the load was uniformly distributed round a concentric circle. Australian mathematician J. H. Michell FRS (1863–1940) investigated the case where the load was concentrated at a point. In his published paper titled '*On a problem of elastic circular plates*' [Bull. Cal. Math. Soc., 16, (1925), 63–70], Prof. S. Ghosh obtained the form of the deflection of a plate when (a) the load was uniformly distributed between two concentric circles and two radii and (b) the load was distributed along the arc of a concentric circle. In another paper published the same year and titled '*On the solution of $\Delta_1{}^4 w = C$ in bipolar coordinates and its application to a problem in elasticity*' [Bull. Cal. Math. Soc., 16, (1925), 117–122], the solution was obtained by him in bipolar co-ordinates and used to find the deflection of an elastic plate bounded by two non-concentric circles bent by its own weight. In the paper entitled '*On the bending of a loaded elliptic plate*' [Bull. Cal.

Math. Soc., 21, (1929), 191–194] Prof. S. Ghosh obtained the solution of bending of an elliptic plate with the load concentrated at the focus.

Prof. S. Ghosh made notable research on the problems of dislocation as they occur following the concept outlined in the mathematical theory of elasticity. According to this theory, there is thus a physical possibility of displacements in multi-connected spaces being expressed by many valued functions. World famous Italian mathematician and physicist V. Volterra FRS (1860–1940) had applied the solutions of equations of equilibrium to the problem of dislocation in a hollow cylinder bounded by two concentric circles in terms of polar co-ordinates. Prof. S. Ghosh in his paper titled *'On certain many valued solutions of the equations of elastic equilibrium and their application to the problem of dislocation in bodies with circular boundaries'* [Bull. Cal. Math. Soc., 17, (1926), 185–194] used the expression for the stress function X in bipolar co-ordinates for solving the problem of dislocation in a hollow cylinder having its cross section bounded by two non-concentric circles due to (a) parallel fissures and (b) wedge-shaped fissures. In the paper entitled *'On the solution of the equations of elastic equilibrium suitable for elliptic boundaries'* [Transactions of American Mathematical Society, 32, (1930), 47], Prof. S. Ghosh solved biharmonic equations using elliptic co-ordinates. He then applied it to the problem of plane strain. He found that the method used by A. E. H. Love (1863–1940) for solving problems in plane strain in elliptic co-ordinates is applicable only to cases where the surface displacements are given. The method developed by Prof. Ghosh in the above-mentioned paper gave in addition to Love's solution the solution of an infinite solid with elliptic cylindrical hollow subjected to a pressure on the inner elliptic boundary. His method also gave solution of the problem of dislocation for a wedge-shaped fissure in a body that was bounded by two confocal elliptic cylinders.

C. Chree FRS (1860–1928) had solved the problem for determining stress and strain in rotating elliptic cylinders and discs by using Cartesian co-ordinates. But the expressions thus obtained were very complicated. N. M. Basu and H. M. Sengupta in their paper titled *'On the strain in a rotating elliptic cylinder'* [Bull. Cal. Math. Soc., 18, (1927), 141–145] had used elliptic co-ordinates; but their method is applicable only to slowly rotating cylinders with nearly circular sections. However, Prof. S. Ghosh in his paper titled 'On the plane strain and stress in rotating elliptic cylinders and discs' [Bull. Cal. Math. Soc., 19, (1928), 117–126] considered the problem in a more general way and solved the same for the stress function X without any limitations. His other important contributions in this area of 'Plane Problems' are listed below:

- *'On the stress and strain in a rolling wheel'* [Bull. Cal. Math. Soc., 25, (1933), 99–106],
- *'Plane strain in an infinite plate with an elliptic hole'* [Bull. Cal. Math. Soc., 28, (1936), 21–47],
- *'Stress distribution in a heavy circular disc held with its plane vertical by a peg at the centre'* [Bull. Cal. Math. Soc., 28, (1936), 145–150],
- *'On the distribution of stress in a semi-infinite plate under the action of a couple at a point in it'* [Bull. Cal. Math. Soc., 29, (1937), 177–184],

- *'Stress distribution in an infinite plate containing two equal circular holes'* [Bull. Cal. Math. Soc., 31, (1939), 149–159],
- *'On plane strain and plane stress in aeolotropic bodies'* [Bull. Cal. Math. Soc., 34, (1942), 157–169],

Prof. S. Ghosh also carried out studies in the areas of 'Vibration of Ring' and 'Torsion and Flexure Problems'. Special mention needs to be made of his three papers listed below:

- *'On the flexure of an isotropic elastic cylinder'* [Bull. Cal. Math. Soc., 39, (1947), 1–14]
- *'On a new function-theoretic method of solving the torsion problem for some boundaries'* [Bull. Cal. Math. Soc., 39, (1947), 107–112]
- *'On the flexure of a beam whose cross-section is bounded partly by a straight line'* [Bull. Cal. Math. Soc., 40, (1948), 77–82].

In these three papers, Prof. S. Ghosh's remarkable mathematical ability is revealed as he develops function-theoretic methods for solving torsion and flexure problems.

As already mentioned, Prof. S. Ghosh was a leading researcher in the field of Elasticity. He was elected a 'Fellow' of the Indian National Science Academy (INSA) in 1951 for his original and outstanding contributions in Applied Mathematics. His capability of reducing complex engineering problems to those of Pure Mathematics was really remarkable. Many of his research papers have been quoted in standard text books on 'Elasticity' and relevant reference books of that discipline. The problems of 'Torsion and Flexure' as solved by Prof. S. Ghosh and his student Prof. D. N. Mitra have been discussed in details in the book entitled '*Resistance des matèriaux, thèorique et expèrimentale*' [(Vol. I), Dunod, Paris, 1954] authored by Robert L' Hermite. Prof. S. Ghosh was a Member of the 'London Mathematical Society' and the 'Calcutta Mathematical Society'.

Many students obtained their Ph.D. degrees working under the supervision of Prof. S. Ghosh. Some of them became quite well-placed in life. Among his students, Prof. D. N. Mitra was a Professor and Head of the Department of Mathematics, IIT, Kharagpur. Prof. B. Karunes was a Professor and Head of the Department of Applied Mechanics, IIT, Delhi; Prof. J. G. Chakrabarty was a Professor at the Department of Applied Mathematics, University of Calcutta; Prof. S. K. Khamrui was a Professor at the Department of Mathematics, Jadavpur University, Calcutta and Dr. (Sm.) Lakshmi Sanyal was Principal, Sarojini Naidu Girls' College, Calcutta.

Prof. S. Ghosh left his ancestral home in 1919 after the death of his mother. He remained a bachelor and led a very frugal life. After retirement, he gave away most of the money that he received from the University to different charitable institutions, such as Sri Ramakrishna Mission, Bharat Sevasram Sangha, Mother Teresa's 'Sisters of Charity', Salvation Army, Red Cross, etc. He made liberal donations during natural calamities and silently helped many needy people. He was a saintly man and led a life of renunciation.

Towards the end of his life, ill health and rising prices of essential commodities forced him to lead a life of penury. But fortunately two of his former lady students Dr.

Lakshmi Sanyal and Dr. Mina Majumder helped him through his difficult times. They behaved like little mothers and nursed the great scholarly noble man with selfless love and care till the very end. Prof. Ghosh died on the 6 May 1976 in Calcutta. He left behind him his golden legacy of righteousness, kindness and profound scholarship.

8.3   Special Honours

He was elected a 'Fellow' of the Indian National Science Academy (INSA) for his original and outstanding contributions in Applied Mathematics.

8.4   Important Milestones in the Life of Professor Suddhodan Ghosh

1896: Born in a wealthy Bengali family of landlords on the 26 July 1896.

1914: Passed the Matriculation Examination with a first division and a high rank in the merit list.

1916: Passed the I.Sc. Examination from the Scottish Church College, Calcutta. He secured the eighth rank in the combined (I.A. and I.Sc.) merit list of the University of Calcutta. He secured the highest marks in Mathematics among all successful candidates.

1918: He stood First with a First class in the B.Sc. Mathematics (Honours) Examination from Presidency College. He was awarded the University Gold Medal for his brilliant performance.

1920: He again stood First with a First class in the M.Sc Examination in Mixed (Applied) Mathematics from the University of Calcutta. He secured about 90% of the total marks. He was awarded the University Gold Medal for his outstanding performance.

1920: He joined the Department of Mixed (Applied) Mathematics as 'Sir Rashbehary Ghosh Research Scholar' under Prof. N. R. Sen for pursuing research in Applied Mathematics.

1925: He won the prestigious 'Premchand Roychand Studentship' of Calcutta University. He was also awarded the 'Sir Asutosh Mookerjee Gold Medal' of Calcutta University for his outstanding research contributions in Mathematics.

1928: He obtained the D.Sc. degree of the Calcutta University for his original research in Applied Mathematics.

1928: He joined the Department of Mixed (Applied) Mathematics, University of Calcutta as a Lecturer.

1951: Elected 'Fellow' of the Indian National Science Academy (INSA).

1958: He joined the Department of Pure Mathematics as a Professor and Head of the Department.

1959: He joined as the 'Sir Rashbehary Ghosh Professor' and Head of the Department of Applied Mathematics of the University of Calcutta.

1962: He retired from the University of Calcutta.

1976: He passed away on the 6 May 1976, in Calcutta.

Chapter 9
Rabindranath Sen (1896–1974)

9.1 Birth, Childhood, Education

Rabindranath Sen was born on 7 January 1896 in Dacca in erstwhile East Bengal. It is now the capital of the present-day Bangladesh. His father Jogendra Nath Sen was an advocate of the Calcutta High Court. He was also a close friend of Sir Asutosh Mookerjee. Jogendra Nath was elected as Member of the 'Senate' and 'Syndicate' of Calcutta University. R. N. Sen's mother Hemanta Kumari Devi came from a well-to-do family and was a religious and kind-hearted person. Jogendra Nath Sen died in a car accident in Calcutta. Because of his father's sudden demise at a comparatively early age, R. N. Sen stayed with his mother in his maternal grandfather's house in Dacca. He had his schooling from 'Dacca Collegiate School' and graduated from 'Dacca College'. However for his Post-Graduate studies, R. N. Sen came down to Calcutta and joined the Pure Mathematics Department of the Calcutta University. In 1920 he obtained his M.A. degree in Pure Mathematics with a First class from the University of Calcutta.

9.2 Higher Studies, Career, Research Contributions

R. N. Sen started his research work in the area of Differential Geometry. This particular type of Geometry is a mathematical discipline that uses the techniques of Differential Calculus and Integral Calculus as well as Linear Algebra for studying the geometric properties of curves and surfaces. R. N. Sen started working on Differential Geometry with the study of simplexes in n-dimensions. His first research paper entitled '*Simplexes in n-dimensions*' was published in the 'Bulletin of the Calcutta Mathematical Society' [Bull. Cal. Math. Soc., Vol. 18, pp. 33–64, 1926].

In 1928, R. N. Sen proceeded to the University of Scotland, Edinburgh, for pursuing further research under the supervision of the well-known mathematician

© The Author(s), under exclusive license to Springer Nature Singapore Pte Ltd. 2022
P. Mukherji, *Notable Modern Indian Mathematicians and Statisticians*,
https://doi.org/10.1007/978-981-19-6132-8_9

Sir E. T. Whitaker. R. N. Sen was the recipient of the prestigious 'Newton Scholarship'. In a short span of only two years he completed his research work. In 1930, he obtained his Ph.D. degree from the Edinburgh University for his thesis entitled 'On the Newer Theories of Space'.

On his return to India from UK in, he initially worked at 'Asutosh College', a local private College in Calcutta, as a Lecturer in Mathematics for a brief period of two years. However, in 1933 he joined his alma mater, the Department of Pure Mathematics in Calcutta University as a Lecturer. In 1954, Dr. Sen took charge as the Head of the Department of Pure Mathematics and also became the 'Hardinge Professor' of Pure Mathematics. He served in that capacity for seven long years. Prof. R. N. Sen served this well-known Department to the best of his ability. He taught in the Post-Graduate class, guided research students, wrote useful text books and personally did brilliant research during his seven years as the 'Hardinge Professor' of Pure Mathematics. He retired from the Department in 1961. Two special achievements which point to his remarkable ability as an administrator need to be specially mentioned. Ever since its inception, the Department of Pure Mathematics was under the control of the 'University College of Arts'. It was due to Prof. R. N. Sen's intense and tireless efforts that the Department was brought under the aegis of the 'University College of Science'. The second important action was the reorganization of the Departmental Library, keeping the requirements of the research activities in mind. It would not be an overstatement to say that Professor R. N. Sen was the luminous star of the Department of Pure Mathematics of the University of Calcutta.

Prof. R. N. Sen made enormous research contributions in the field of Differential Geometry. His research activities may broadly be classified as follows:

1. Differential Geometry of Riemannian Spaces. In this context it may be noted that Riemannian Geometry is a Non-Euclidean Geometry in which straight lines are geodesics and in which Euclid's parallel postulate is replaced by the postulate that within a plane every pair of lines intersects. A classical Riemannian manifold has a metric defined by a special Riemannian tensor.
2. Differential Geometry of Finsler Spaces. Incidentally, Finsler Geometry is just Riemannian Geometry without the quadratic restriction. It derived its name from the thesis submitted by Finsler in 1918. Actually Finsler Geometry has the Finsler manifold as the main object of study.

After completing his Ph.D. degree for sometime Dr. R. N. Sen worked on problems related to hypersurfaces. His two notable papers on this area are *'On Curvature of a Hypersurface'* [Bull. Cal. Math. Soc., 23, (1931), 1–10] and *'On Rotations in Hypersurfaces'* [Bull. Cal. Math. Soc., 23, (1931), 195–209]. In 1931, in the 'Proceedings of the Edinburgh Mathematical Society', Dr. R. N. Sen published a research paper entitled *'On One Connection between Levi-Civita parallelism and Einstein's Teleparallelism'* [Proc. Edinb. Math. Soc., Ser. 2, Part 4, (1931), 252–255]. It may be noted that 'teleparallelism' means an arbitrary parallelism known as distant parallelism. Between 1931 and 1946, he published a number of research papers on parallelism including a series of papers on parallelism in Riemannian Space.

His work connected with teleparallelism received wide acclaim from internationally well-known geometrician Professor T. Levi-Civita of Rome University, Italy.

Prof. M. C. Chaki, the most famous student of Professor R. N. Sen, has stated: 'Sen's investigations on the behaviour of an arbitrary parallel displacement in a metric space resulted in the discovery in 1949–1950, of an algebraic system of affine connections in which Levi-Civita parallelism could be identified. This work is considered as highly significant and is referred to by I. M. H. Etherington of Edinburgh University as 'Senian Geometry'. The algebraic system of affine connection which Prof. R. N. Sen discovered was applied to Finsler Spaces. The four papers related to an algebraic system generated by a single element and its application in Riemannian Geometry are listed below:

- '*On an Algebraic System Generated by a Single Element and its Application in Riemannian Geometry*' [Bull. Cal. Math. Soc., 42, (1950), 1–13],
- '*On an Algebraic System Generated by a Single Element and its Application in Riemannian Geometry II*' [Bull. Cal. Math. Soc., 42, (1950), 117–187],
- '*On an Algebraic System Generated by a Single Element and its Application in Riemannian Geometry III*' [Bull. Cal. Math. Soc., 43, (1951), 77–94],
- '*Corrections to my papers on an Algebraic System* etc.' [Bull. Cal. Math. Soc., 44, No. 2, (1952)].

Published between 1950 and 1952 these 4 research papers are all important contributions of Prof. R. N. Sen. Special mentions need to be made of Professor R. N. Sen's following two research papers:

- '*On New Theories of Space in General Relativity*' [Bull. Cal. Math. Soc., 56, (1964), 1–14],
- '*On New Theories of Space in Unified Field Theory*' [Bull. Cal. Math. Soc., 56, (1964), 147–162].

As is clear from the titles of the above-mentioned papers, they deal with general relativity and unified field theory. These research papers were highly praised by famous mathematicians and physicists. According to Prof. M. C. Chaki, this work has been cited in the following important books:

1. 'Theory of Linear Connections' by Dr. J. Strik
2. 'Ricci Calculus' by J. A. Sohontin
3. 'Vorlesungen uber Diffusion Geometry' by G. Vranceanu.

This work of Prof. R. N. Sen has been mentioned and discussed in the well-known book titled 'On the History of Unified Field Theories' [Springer] written by Hubert F. M. Goenner of Göttingen University. This work has also been cited by various researchers in India and abroad.

Prof. Ram Bilas Misra in his book entitled 'Differential Geometry, its past, present and future' has extensively discussed Prof. R. N. Sen's contributions in different areas of Differential Geometry.

Many students of Calcutta University obtained their Ph.D. degrees working under the guidance of Prof. R. N. Sen. One of his students Dr. Hrishikesh Sen worked and later became a Faculty Member of Burdwan University, West Bengal. As already mentioned M. C. Chaki was Prof. R. N. Sen's most successful and famous student. Prof. Chaki also worked on Finsler Geometry and on various other branches of geometry. He was a teacher of eminence and a very versatile mathematician.

Prof. R. N. Sen received many academic and professional honours in his illustrious career. To mention a few, he was elected a Fellow of the National Academy of Sciences (now known as Indian National Science Academy) in 1954, President of the Mathematics Section of the Indian Science Congress in 1956 and President of Calcutta Mathematical Society for three successive terms in 1963–1966.

A distinguished academician, a capable and innovative administrator, a brilliant teacher and a very modest and humane human being, Professor Rabindranath Sen was a luminary among the mathematicians of Bengal. He enriched mathematical teaching and research in India with his selfless and dedicated service to the noble cause. He breathed his last in Calcutta on 19 July 1974.

9.3 Important Milestones in the Life of Professor Rabindra Nath Sen

1896: Born on 7 January 1896 in Dacca, capital of present-day Bangladesh.

1920: Obtained the M.A. degree in Pure Mathematics with a First class from the University of Calcutta.

1928: Proceeded to Scotland, UK, to pursue higher studies with 'Newton Scholarship'.

1930: Received Ph.D. from the University of Edinburgh, Scotland, for his thesis titled 'On the Newer Theories of Space'.

1933: Joined the Department of Pure Mathematics, Calcutta University, as a Lecturer in Mathematics.

1954: Became 'Hardinge Professor' and Head of the Department of Pure Mathematics, Calcutta University.

1954: Elected Fellow of the National Academy of Sciences, India.

1956: Elected President of the Mathematics Section, Indian Science Congress,

1961: Retired from the University of Calcutta.

1963–1966: Elected President, Calcutta Mathematical Society for 3 successive terms.

1974: Died on 19 July 1974 in Calcutta.

Chapter 10
Bibhuti Bhusan Sen (1898–1976)

10.1 Birth, Education

Bibhuti Bhusan Sen was born on 1 March 1898 in a village in the district of Chittagong in the erstwhile East Bengal (present-day Bangladesh). His father Shyama Charan Sen was a government pleader, and his mother Dusmanta Kumari Sen was a kind-hearted, pious housewife. B. Sen was a brilliant student and showed signs of brilliance right from his school days. He passed the Matriculation Examination then conducted by the University of Calcutta in 1913. He stood first from the district of Chittagong and obtained a Government Scholarship. After that he came to Calcutta and joined the famous Presidency College in the intermediate class. In 1915 he passed the Intermediate Examination in Science with a first division. He won scholarships throughout his academic career. In 1917 he passed the B.Sc. (Honours) Examination in Mathematics standing First with a First class. Though B. Sen got himself enrolled for the M.Sc. course in Mixed Mathematics at the University College of Science in Calcutta University, due to some problems at home, he could not sit for the Final Examination in 1919. However, he got over the temporary setback through his perseverance and intelligence and completed the M.Sc. Examination in 1921. He not only stood First with a First class but scored highest marks among all successful candidates in the M.Sc. Examination in all subjects. He was awarded the 'University Gold Medal' for his outstanding performance in the Examination.

10.2 Career, Research Contributions

After completing Post-Graduation, B. Sen had brief stints of teaching. He first taught in a private college in present-day Bangladesh and then in the Ramjash College in Agra, UP. After spending a year in Agra, he returned to Calcutta. There are reports that Sir Asutosh Mookerjee, the then Vice-Chancellor of Calcutta University had wished to bring in B. Sen as a full-time Lecturer in the Department of Mixed

P. Mukherji, *Notable Modern Indian Mathematicians and Statisticians*,
https://doi.org/10.1007/978-981-19-6132-8_10

Mathematics in the University College of Science of Calcutta University. But due to sudden and untimely demise of Sir Asutosh (in Patna), B. Sen was deprived of this privilege. In 1924, he joined the 'Bengal Educational Service'. His first posting was at Chittagong Government College. Later on he was transferred to Islamia College (now known as Maulana Azad College), Calcutta. He served there till 1933. He was then transferred to Krishnanagar College and worked there till 1939. After that in 1940 he was again transferred back to Islamia College. In 1943, B. Sen was posted at the Bengal Engineering College at Shibpur, Howrah. He joined the famous Presidency College in Calcutta in 1945. Then in the same year he was again transferred back to the Bengal Engineering College and served there till 1948. In 1949 he was again brought back to the Presidency College and served there till 1951. He is remembered as an outstanding Faculty Member of this prestigious college. Finally he was made the Principal of the Hooghly Mohsin College in 1952, and he retired from there in 1953 at the age of 55, as per rules of those times. He thereafter went to Birla College of Science as a Professor of Mathematics. After a brief stint there, in 1956, at the request of Dr. Triguna Sen, the then Rector Of Jadavpur University in Calcutta, Dr. B. Sen joined the Mathematics Department of the said University as a Professor and Head of the Department. Under his dynamic leadership started the undergraduate and Post-Graduate courses in Jadavpur University. Simultaneously, Prof. B. Sen put in a lot of hard work and effort to gradually build up a strong school of research in Continuum Mechanics there.

During his student days, in the Post-Graduate class of the Mixed Mathematics (Applied Mathematics) Department, B. Sen had the good fortune of being taught by stalwart applied mathematicians like Prof. S. N. Bose, Prof. N. R. Sen and some others. His special paper on the 'Theory of Elasticity' was taught by Prof. S. N. Bose. B. Sen was enamoured by his marvellous teaching technique and gradually developed a great liking for the subject. In his heart B. Sen had also probably a desire to take up research in the discipline. But in the early twentieth-century India there was hardly any proper infrastructure available for even carrying out research on theoretically oriented subjects like Mathematics. There were no books on higher Mathematics written by Indian authors available in the country. All books were written by foreign mathematicians and published abroad. So they were pretty expensive and far beyond the reach of students from middle class homes. Photocopy technology was unheard of, and typing facilities were also scarce in India. Even the newly established Department of Mixed Mathematics (Applied Mathematics) could not provide their students with necessary books and journals. The colonial rulers were not interested in allocating adequate funds to the Universities, for such expenses. But students like B. Sen had indomitable spirits and determination to conquer such adversities. Since he could not buy the books due to paucity of funds, so B. Sen personally copied by hand the two voluminous books namely '*A Course of Modern Analysis*' by E. T. Whitaker and G. N. Watson and '*A Treatise on the Mathematical Theory of Elasticity*' by A. E. H. Love. Frequent transfers to remote places and repeated posting in degree colleges could not dampen the enthusiasm of this mathematical genius from achieving excellence in his chosen field. Just to show how committed he was to his cause, it may be sufficient to point out from available records that between 1933 and 1939, during his posting at Krishnanagar College, B. Sen had published as many as 19 research papers. He had no supervisor to guide him, no suitable library facilities, nor necessary books and journals, and yet 6 of those 19 papers were published in the famous British Journal '*Philosophical Magazine*' and 1 written in German and published in the well-known German journal ZTP [*Z fur Tech Phys,*]. This trend continued throughout his service period. It never mattered to him, where he was. He continued making valuable contributions in various branches of Elasticity. B. Sen carried out notable research work in the areas of Elastodynamics, Thermo-elasticity, Visco-elasticity, Magneto-elasticity, Piezo-elasticity, Plasticity, Couple Stresses, Theoretical Seismology, etc. He perhaps guided the largest number of Ph.D. students (about 50) among all the contemporary mathematicians of his time. Apart from Elasticity, he also did commendable research work on Fluid Mechanics. He supervised some Ph.D. students to solve problems related to Fluid Mechanics as well.

Though Prof. B. Sen was regularly publishing high-quality research papers in well-known, high-end journals, but perhaps he was never very keen to acquire a formal research degree himself. He finally submitted a hand-written [not typed] thesis and earned the D.Sc. degree from the University of Calcutta. By that time 40 of his earlier research papers had already been published in renowned journals. The only other known famous Indian mathematician who had submitted a hand-written thesis and acquired a Ph.D. degree was Hans Raj Gupta of Panjab University. Prof. B. Sen did commendable research in various branches of Elasticity, and his research work has been widely discussed in famous books such as *Thermoelasticity* by Witold Nowacki [Pergamon Press, New York, (1962)], '*Theory of Elasticity*' by Stephen Timoshenko and James N. Goodier [McGraw Hill, New York, (1951)], *Theory of Plates and Shells* by Stephen Timoshenko and Winowsky-Kreiger [McGraw Hill, New York (1959)].

Prof. B. Sen had published 57 research papers on various topics of continuum mechanics in reputed national and international journals. As already mentioned earlier, Prof. B. Sen was a prolific researcher in the field of Elasticity. Two of his publications in this field deserve to be specially noted. They are:

- '*Two dimensional boundary value problems of Elasticity*' [Proc. Royal Soc. London, A, 187, (1946), 87–101]
- '*Note on the direct determination of steady state thermal stresses in circular discs and spheres*' [Bull. Calcutta Math. Soc., Sir Asutosh Mookerjee Birth Centenary Commemoration Volume, 56, (1964), 77–81].

Prof. B. Sen's contributions in the field of Fluid Mechanics are also commendable. In this context special mention of the following two research papers is essential. They are:

- '*Note on the application of trillinear coordinates in some problems of elasticity and hydrodynamics*' [Bull. Calcutta Math. Soc., 27, (1935), 73–78]
- '*Note on the flow of viscous liquid through a channel of equilateral triangular section under exponential pressure gradient*' [Rev. Roumania Sci. Tech. Ser. Mecanique Appl., 9, (1964), 307–310].

It may be noted that, as a matter of principle, he never claimed joint-authorship of research papers published by his Ph.D. students. Some of his Ph.D. students became successful professionals in later life. Prof. Manindra Nath Mitra, Prof. Arabinda Mukhopadhyay, Prof. Achintya Kumar Mitra, Prof. Bikash Ranjan Das, Prof. Kripa Sindhu Chaudhury and many others who had been Ph.D. students under the supervision of Prof. B. Sen kept his flag flying by becoming good teachers and researchers themselves. They all became Professors in various notable Universities and institutions. They also supervised many Ph.D. students successfully.

Prof. B. Sen had also authored three books on Higher Mathematics, which are listed below:

1. *A Treatise on Special Functions* [Allied Publishers, Calcutta (1967)]
2. *An Elementary Treatise on Laplace Transforms* [The World Press Pvt. Ltd., Calcutta, (1969)]
3. *A Treatise on Numerical Analysis* [Dasgupta and Co., Calcutta, (1977)].

The last book mentioned above was published after the death of Prof. B. Sen.

10.3 Special Honours

The 1952 was a special year in the life of Prof. B. Sen. That year he was elected a Fellow of the National Institute of Sciences [presently known as Indian National Science Academy (INSA)]. That same year he was also elected President, Mathematics Section of the Indian Science Congress Association. He was also elected the 'President' of the Calcutta Mathematical Society.

Prof. B. Sen's fame as a brilliant mathematician and research guide was so great, that in 1963, the then Prime Minister of India, Sri Jawaharlal Nehru, personally intervened and persuaded UGC to give special permission, so that Prof. B. Sen who had already crossed the age of 65 could join the Visva Bharati University. In 1963, Prof. B. Sen joined the historic Visva Bharati University at Shantiniketan, which had been founded by Nobel Laureate poet Rabindra Nath Thakur. Prof. B. Sen took charge as the Professor and Head of the Department of Mathematics, which had started functioning from 1961. Prof. B. Sen virtually built up the new Department of this Heritage University. As was his wont he started building a school of research on continuum mechanics in the new department. He later on adorned the chair of 'Principal' of 'Vidya Bhabana' (College of Post-Graduate Studies) at Visva Bharati.

In the context of Visva Bharati, to mention a small interaction of Prof. B. Sen with the poet may not be out of place. Years earlier after reading the poet's famous science-related book written in Bengali '*Viswaparichaya*' Prof. B. Sen wrote a letter to the Nobel Laureate poet Rabindranath Thakur suggesting some necessary corrections. The poet not only incorporated the corrections but duly acknowledged Prof. Sen's suggestions in the 'Preface' of the next edition.

Finally in 1968 he retired from there and returned to Calcutta. Jadavpur University honoured Prof. B. Sen as an Emeritus Professor in the same year. He continued in that capacity till he breathed his last on 13 December 1976.

A brilliant mathematician, a superlative research guide and a caring teacher, he was a key figure in the development of mathematical research in Bengal during the twentieth century.

10.4 Important Milestones in the Life of Professor Bibhuti Bhusan Sen

1898: Born on 1 March 1898 in the district of Chittagong in undivided Bengal.

1913: Passed the Matriculation Examination, standing first in the district of Chittagong.

1915: Passed the Intermediate Examination in Science from Presidency College, Calcutta, in first division and won a merit scholarship.

1917: Passed the B.Sc. (Honours) in Mathematics standing First with a First class from the University of Calcutta.

1921: Passed the M.Sc. Examination in Mixed Mathematics (Applied Mathematics) standing First with a First class from the University of Calcutta. He was the recipient of the 'University Gold Medal'.

1924: Joined the 'Bengal Educational Service'.

1946: Obtained the D.Sc. degree in Mathematics from the University of Calcutta.

1952: Elected 'Fellow' of the National Institute of Sciences. Elected 'President' of the Mathematics Section, Indian Science Congress held at Calcutta.

1952: Joined as Principal of Hooghly Mohsin College, West Bengal.

1953: Retired from 'Bengal Educational Service'.

1956: Joined Jadavpur University, Calcutta, as Professor and Head of the Department of Mathematics.

1963: Joined the Visva Bharati University, Shantiniketan, West Bengal, as Professor and Head of the Department of Mathematics.

1968: Became 'Emeritus Professor' at Jadavpur University.

1976: Breathed his last on 13 December 1976 in Calcutta.

Chapter 11
Raj Chandra Bose (1901–1987)

11.1 Childhood, Family, Education

Born on 19 June 1901 in Hoshangabad, Madhya Pradesh, in a cultured Bengali family, Raj Chandra was the eldest son of Pratap Chandra Bose, an army doctor, and Ushangini Devi, with three younger siblings. His early life was spent in Rohtak, a small town in Punjab (now in Haryana). It was recognized at a very early age that he was a very intelligent child with a photographic memory, and not surprisingly this led the parents, especially his father, to have high ambitions about his future. As a result, while he had a happy childhood, he was often on tenterhooks to perform academically. This was compounded by his having a rather delicate constitution. Notably, during his Matriculation Examination he was afflicted by influenza, could not fare very well in the examination, and as a result he marginally missed the merit scholarship.

In order to continue his studies, he had to move to Delhi in 1917 as without a scholarship it was not possible for him to carry on his studies in Punjab. In 1919, the year of the Jalianwalla Bag massacre, the Intermediate Examinations of Punjab University were postponed due to political unrest. When the examinations were finally held, he stood first in the Intermediate Examination, much to the satisfaction of his father. The situation, however, took a turn for the worse shortly afterwards, when he lost his father; he had lost his mother earlier. This led to dire financial crisis for the family, now consisting only of the siblings, and Raj Chandra was constrained to take up a job in a secondary school and also to give some private tuitions to make both ends meet. Reflecting on the immense hardship that they had to undergo during the period, Bose once wrote: ['*Autobiography of a mathematical statistician; The making of Statisticians*', J. Gani (Ed.), Springer-Verlag, New York (1982)].

In the summer of 1921, after my brother had completed school, I moved my family to Delhi, where the four of us lived in a single room provided by the family of a student I was coaching. My brother started college and took a coaching job. With this income and the scholarships which both of us were getting, we somehow managed to survive. My B. A. Honours examination (in 1922) was two months away when our coaching jobs expired. My scholarship was also due to expire. We were on the brink of starvation on the eve of the examination.

© The Author(s), under exclusive license to Springer Nature Singapore Pte Ltd. 2022
P. Mukherji, *Notable Modern Indian Mathematicians and Statisticians*,
https://doi.org/10.1007/978-981-19-6132-8_11

Fortunately, we could sell a bottle of quinine from the stock of medicines left by my father.
This fetched a good price as quinine was very scarce in the aftermath of the First World War

11.2   Higher Education and Research

Raj Chandra Bose continued his education in Delhi, alongside fighting relentlessly against poverty, and shouldering the burden of bringing up his younger brother and two younger sisters. In 1924, he obtained the M.A. degree in Applied Mathematics from the University of Delhi (the Delhi University had been carved out of the Punjab University in 1921). Although he secured highest marks among the Post-Graduate students, he was not awarded first class due to shortage in attendance in Post-Graduate classes as per the requirement of the University. His effort to get a teaching assignment in Delhi was not successful. Seth Kedarnath Goenka, a business man of Delhi, attracted by his teaching abilities, promised to support him to continue his studies in Calcutta (now Kolkata) which he was aspiring. About this transition Raj Chandra reminisces. [*'Autobiography of a mathematical statistician; The making of Statisticians'*, J. Gani (Ed.), Springer-Verlag, New York (1982)].

> *Now my fortunes took another turn. Seth Kedarnath Goenka, whose younger brother I was coaching, was so much pleased with my work that he agreed to support me as long as I needed, in Calcutta, where I wanted to study Pure Mathematics. I thus came to Calcutta in 1925............... . Here I had the good luck to catch the attention of Prof. Syamadas Mukherjee, a fine geometer of the old school. His main interests at that time were non-Euclidean geometry, n-dimensional geometry and global (imgrössen) properties of convex curves. He gave me a room in his house and free use of his library. He also secured a good coaching job for me so that I no longer needed any help from Sethji.*

Thus began the Calcutta years of Raj Chandra as a Post-Graduate student in the Department of Pure Mathematics of Calcutta University, and in 1927 he stood First with a First class in the M.A. Examination in Pure Mathematics at the University of Calcutta. He won the University Medal and the K. Mallik Gold Medal, which were a laudable achievement, especially as the Calcutta University at that time was the epicentre of mathematical teaching and research in the country, and the best students from all over India competed for the honours.

Raj Chandra Bose was a research associate in Mathematics from January 1928 to December 1929 in the Department of Pure Mathematics and worked under the guidance of Professor Syamadas Mukhopadhyay. In fact he also stayed at the latter's house. Acknowledging Professor Syamadas Mukhopadhyay's contributions in his life, Raj Chandra Bose wrote: [*'Autobiography of a mathematical statistician; The making of Statisticians'*, J. Gani (Ed.), Springer-Verlag, New York (1982)].

> *He was a dedicated teacher; from him I learned the spirit of mathematical research. I also made good use of his library and learned all the mathematics I could. My special favourites were Hilbert's Grundlagen der Geometrie and Klein's Icosahedron and Elementary Mathematics from a Higher Point of view. The stimulus thus provided has had a profound influence on my mathematical career.*

Facilities for research and academic activities were very limited in India during the pre-independence period. It was with great difficulty that R. C. Bose managed to get a job in Asutosh College, an undergraduate college in Calcutta. In spite of the heavy workload at the College, Raj Chandra pursued his studies in the field of Geometry under the guidance of and inspiration from his former teacher and mentor Professor Syamadas Mukhopadhyay. His first research paper on Hyperbolic Geometry, in collaboration with Professor Mukhopadhyay, was published in 1926. He never looked back after that. He applied himself with single-minded devotion to the topic, and in about the next seven years until 1933, he had over a dozen original research papers of high quality to his credit, in the areas of Differential Geometry of Convex Curves and Hyperbolic Geometry.

In 1931 Professor P. C. Mahalanobis founded the Indian Statistical Institute (ISI), which initially operated from a small room in the Presidency College in Calcutta. He was in search of suitable people to advance the study and research of Statistics in India. Attracted by Bose's prolific research work, Professor P. C. Mahalanobis personally went to meet Bose at his residence in 1932 and invited him to join as part-time Member of the newly founded institute. Bose, after some hesitation, accepted his offer. He worked for the institute on Saturdays and during vacations of the college. In January 1933, R. C. Bose joined the ISI as a Research Scholar. Though he hardly had any background in Statistics, he evidently had great potential, in the eyes of Professor P. C. Mahalanobis.

Regarding his initiation into Statistics, Bose wrote: ['*Autobiography of a mathematical statistician; The making of Statisticians*', J. Gani (Ed.), Springer-Verlag, New York (1982)].

> .... *The Secretary brought me the volume of Biometrika from (1900-1932), the most important statistical journal at that time, together with a typed list of 50 papers. There was also R A Fisher's book Statistical Methods for Research Workers. When I went to see Mahalanobis, he said 'you were saying that you do not know much Statistics. You master the 50 papers, the list of which you have received, and Fisher's book. This will suffice for your statistical education for the present'. Thus began my education in statistics.*

A few months later Professor Mahalanobis also recruited another young and brilliant mathematician, S. N. Roy, a product of the department of Applied Mathematics of Calcutta University. R. C. Bose and S. N. Roy became collaborators and close friends and took charge of advancing mathematical statistics at ISI. In 1935, both R. C. Bose and S. N. Roy were made regular appointees of ISI.

Interestingly, within a year of his turning to Statistics, R. C. Bose contributed a paper in the subject, bringing in his expertise in Hyperbolic Geometry to the newly adopted area ['*On the application of hyperspace geometry to the theory of multiple correlation*', Sankhya, Vol. 1, pp. 338–342, 1934].

It may be recalled that in 1925, P.C. Mahalanobis had introduced the novel concept of a measure of divergence between two populations. This idea was further developed by him, on the theoretical side, in a paper in 1928. In the latter paper he found the first four moments of what is called the classical form of D^2-statistic, for the uncorrelated case. He derived certain approximate results for the 'studentised' form of the statistic. In 1935, R. C. Bose obtained the exact distribution as well as the moment of the classical D^2-statistic for the correlated case. This work was written as a research paper titled 'On the exact distribution and moment coefficients of the D^2-statistic' and was published in Sankhya in 1936.

Subsequently, R. C. Bose also published, in collaboration with S. N. Roy, another paper entitled '*Distribution of Studentised D^2-statistic*' in Sankhya in 1938 [Vol. 4, Part I, Special Conference Number, pages 19–38]. At the end of the paper there is a record under the heading '*Discussion on R. C. Bose and S. N. Roy's Paper*', relating to the conference presentation, and the following appreciative comment of R. A. Fischer, a doyen in the area of that time, would be worth recalling:

> *Professor Fisher remarked that he and Professors Hotelling and Mahalanobis had been unwittingly treading the same ground. He was glad to avail himself of the present opportunity to clear up this point. Messers R. C. Bose and S. N. Roy's paper has carried the work a distinct step forward.*

Along with the above, Professor Mahalanobis had noted that the paper had supplied the necessary mathematical tool to use the D^2-statistic when only the sample values of dispersion were known. This paper is considered to be R. C. Bose's greatest contribution in mathematical statistics.

Using R. A. Fisher's concept of representing an n-sample by a point in a Euclidean space of dimension n, P. C. Mahalanobis, S. N. Roy and R. C. Bose introduced the idea of t-coordinates, or 'rectangular coordinates'. This result was published as 'Normalisation of variates and the use of rectangular coordinates in theory of sampling distributions' in Sankhya in 1936.

There took place an important development at that time which had a profound influence on the research career of R. C. Bose. Around 1936, the famous German mathematician F. W. Levi, had to leave Germany because of the anti-Jewish policies of the Nazi rulers. Calcutta University was extremely fortunate in getting him, appointing him as the 'Hardinge Professor' and Head of the Department of Pure Mathematics, which was lying vacant after the demise of Professor Ganesh Prasad. Various new developments were taking place in Europe in Pure Mathematics and Levy introduced them into the syllabus of Calcutta University. Thus the topics in Abstract Algebra like groups, rings, fields, new discoveries in Projective Geometry, Non-Euclidean Geometry, the applications of Vector Geometry and Vector Calculus were all taught at the University providing the students exposure to the topics at par with their international counterparts. Levy also used to organize seminars on Algebra and Geometry. Bose used to attend regularly those seminars lectured by Levi and also had a chance to interact freely with Levy. He became especially interested in finite fields and finite geometries, and in all probability this helped him develop his thought process in the modern style. Moreover in 1938, Levy offered him a half time position in the University of Calcutta to teach Modern Algebra, as it was referred to those days in the Department of Pure Mathematics.

Another striking development was to follow. In Bose's own words: ['*Autobiography of a mathematical statistician; The making of Statisticians*', J. Gani (Ed.), Springer-Verlag, New York (1982)].

'R. A. Fisher, the leading statistician in the world at that time, came in 1938–1939 as a visiting Professor to the Institute. At that time he was interested in the study and construction of patterns called statistical designs, to be used in statistically controlled experiments. He posed to us some problems arising out of this study. It occurred to me that I could successfully use the theory of finite fields and finite geometries, which I was then teaching, as tools for the construction of designs. I solved during his stay one of the problems he had posed. He asked me to develop my methods systematically and send him a paper which he promised to publish in the *Annals of Eugenics* of which he was the editor. Thus my paper on '*The Construction of Balanced Incomplete Designs*' came to be written in 1939; it has now become a classic and is quoted in every book on the subject.

The fortuitous combination of circumstances described above changed the direction of my work and I was to devote the next 18 years of my life (1940–1958) almost wholly to the study of various aspects of statistical designs.'

R. C. Bose was arguably a pioneer in application of finite geometries, finite fields and the combinatorial properties in the construction of designs. His 1938 paper 'On the application of the properties of Galois fields to the problem of construction of hyper Graeco-Latin Squares' [Sankhya, Vol. 3, 1938, pp. 323–339] is very significant. In the introduction Bose had written:

It is hoped that the properties of Galois fields and the finite geometries connected with them, will prove useful in many problems of experimental design and the author hopes to pursue this matter in subsequent papers.

It may be recalled here for the benefit of a general reader that a field is a structure on a set, incorporating operations of addition, multiplication, subtraction and division, satisfying certain basic rules that one is familiar with in the case of rational numbers (fractions) or real numbers. A finite field is one where the underline set is finite (unlike in the case of real numbers). Such a field is also known as Galois field (so named in honour of Evariste Galois). Indeed until the middle of 1950, Bose was mainly pre-occupied with developing a mathematical theory of designs using the geometries associated with spaces modelled over finite fields.

Following relentless persuasion by Professor P. C. Mahalanobis, the University of Calcutta finally decided to start a separate Department of Statistics, and in 1941 the newly set up department started, with Professor P. C. Mahalanobis as the Honorary Head, with R. C. Bose and S. N. Roy as part-time appointees. For sometime R. C. Bose continued to teach both in the Department of Pure Mathematics as well as in the Department of Statistics, but after a couple of years, he resigned from the former and became a full-time teacher in the Department of Statistics. In 1945, when Professor P. C. Mahalanobis relinquished the Headship on account of his commitments at ISI, R. C. Bose took over the Headship of the Department. In 1947 he obtained the D. Lit. degree from the Calcutta University for his original work on Multivariate Analysis and Design of Experiments. One of the examiners of the thesis was R. A. Fischer.

This is the historical account of Professor R. C. Bose's academic activities during his stay in Bengal. But just for the sake of completeness, a few things need to be mentioned. Shortly after obtaining his doctorate degree, R. C. Bose went to the USA and worked there as an invited Professor in several famous Universities. Ultimately he permanently moved over to USA and worked for a long time at the University of North Carolina. After retiring from there in 1971 he joined the Colorado State University, Fort Collins, and worked there till 1980.

While in USA, R. C. Bose continued to work on diverse topics, notable among them being PBIB designs, mutually orthogonal Latin squares, factorial designs, rotatable designs, multidimensional designs, coding theory, file organization, graph theory, additive number theory, projective geometry and partial geometries.

The following are some of the themes with which the name of Professor R. C. Bose is indelibly associated:

- Differential Geometry of Convex Curves and Hyperbolic Geometry
- Association schemes
- Bose–Mesner Algebra
- The mathematical theory of design of experiments using statistical and mathematical methods
- Coding theory including error-correcting codes
- Connection between graph theory and designs.

Some of R. C. Bose's famous student collaborators include C. R. Rao, S. S. Shrikhande, D. K. Roy Chaudhuri, K. R. Nair. The famous number theorist S. Chowla also collaborated with him.

In his research carried out during his stay at USA, two very remarkable results need to be specially mentioned. In three papers jointly written with S. S. Shrikhande and E. T. Parker, Prof. R. C. Bose nullified Euler's Conjecture on the non-existence of mutually orthogonal Latin squares of order twice an odd number. This earned him the nickname of 'Euler-spoiler'.

The second remarkable achievement was the development of an efficient system of error-correcting binary codes known as BCH codes jointly with D. K. Roy Chaudhury.

11.3   Special Honours

Professor R. C. Bose was bestowed with many awards from India as well as the USA. The Indian Statistical Institute, Calcutta, conferred on him D.Sc. (*Honoris Causa*) in 1971. In 1976, he was elected a Member of the prestigious US National Academy of Sciences. In 1979, the Visva Bharati University, founded by Rabindra Nath Tagore, conferred on him the '*Deshikottama*' award. Recently, the Indian Statistical Institute, Kolkata, has with financial support from the central government, set up an advanced Institute of Cryptography. This has been named the '*R. C. Bose Centre for Cryptography and Security*'.

Professor Raj Chandra Bose was an inspiring teacher. Many of his students later became world renowned mathematicians and statisticians. He was a well-read man with a flair for languages and could recite verses in Arabic, Bengali, Persian, Sanskrit and Urdu. He was also a much travelled person. He was an avid gardener and had a keen interest in art, culture and history. He was a gentle and kind person and a nice human being.

Professor R. C. Bose was a great mathematician and an equally great theoretical statistician. His life is a chronicle of a great intellectual endeavour, fulfilled to perfection. He breathed his last in Colorado, USA, in 1987. He will be remembered for long for his outstanding contributions.

11.4 Important Milestones in the Life of Professor Raj Chandra Bose (till 1949)

1901: Born on 29 June 1901 in Hoshangabad of Madhya Pradesh.

1917: Passed the Matriculation Examination from Punjab.

1919: Passed the Intermediate Examination, standing first in Delhi University.

1924: Obtained his M.A. degree in Applied Mathematics from the Delhi University. He stood First with a First class; but the University did not declare it officially, because he did not have the required percentage of attendance in the class.

1927: Stood First with a First class in the M.A. Examination (Pure Mathematics) of the Calcutta University. Won the University Gold Medal and the K. Mallik Medal.

1928–1929: Worked as a Research Scholar in the Pure Mathematics Department of the Calcutta University, under the supervision of Professor Syamadas Mukhopadhyay.

1932: At the personal invitation of Professor P. C. Mahalanobis, he joined the Indian Statistical Institute (ISI), then functioning from a small room in the Baker Laboratory of the Presidency College, Calcutta, as a part-time Research Scholar.

1933: Formally joined as a full-time Research Scholar at the ISI, Calcutta.

1934: Published his famous paper titled '*On the application of hyperspace geometry to the theory of multiple correlation*' [Sankhya, Vol. 4, (1934), 338–342.]

1936: Came in contact with Professor F. W. Levi of the Pure Mathematics Department, Calcutta University.

1938: Came in contact with the world-famous statistician Professor R. A. Fisher, who came to deliver a course of lectures at ISI, Calcutta. Discussions with him inspired R. C. Bose to take up research on 'design of experiments'.

1938: Published the famous research paper titled '*On the application of the properties of Galois fields to the problem of construction of hyper Graeco-Latic Squares*' [Sankhya, Vol. 3, (1938), 323–339.]. R. C. Bose was a pioneer in using finite geometries, finite fields and the combinatorial problems in the construction of designs.

1938: In collaboration with S. N. Roy he published the paper titled '*Distribution of the Studentized D^2 statistic*' [Sankhya, Vol. 4, (1938), 19–38]. This is considered to be R. C. Bose's greatest contribution in 'Mathematical Statistics'.

1939: He published the research paper titled '*On the construction of balanced incomplete block designs*' [Annals Eugenics, (London), Vol. 9, (1939), 358–398]. This paper is considered to be a classic and is referred to in all modern-day text books on the subject.

1941: Joined the newly established Department of Statistics in the University of Calcutta as a part-time teacher.

1945: Was made the Head of the Department of Statistics, Calcutta University.

1947: Obtained the D. Lit. degree from the University of Calcutta for his original work on multivariate analysis and design of experiments.

1949: Professor R. C. Bose permanently emigrated to the USA.

1987: Passed away in the USA.

Chapter 12
Bhoj Raj Seth (1907–1979)

12.1 Birth, Childhood, Education

Bhoj Raj Seth (B. R. Seth) was born at Bhura in the state of Panjab on the 27th August 1907. His family was a believer in high ideals of community service, and these values were inculcated in the mind of B. R. Seth right from his childhood. All his life he did his duty to his family and stood by his students and friends. After completing early education, he moved over to Delhi and joined the Hindu College to continue further studies. In 1927 he obtained the B. A. degree in Mathematics (Honours) standing First with a First class. He was awarded the 'University Gold Medal' for his brilliant performance. In 1929 he passed the M. A. Examination in Mathematics from the Delhi University again standing First with a First class. He once again received the 'University Gold Medal' for his outstanding scholastic ability. As a result of his extra-ordinary performance at the Post-Graduate Examination, he was awarded the 'Central Government Scholarship' for pursuing higher studies in UK. He joined the University of London and enrolled for the M. Sc. course. He studied Elasticity, Photo-Elasticity, Fluid Mechanics and Relativity there. In 1932, B. R. Seth was awarded the M. Sc. degree of the University of London. Thereafter, he started doing research under the guidance of Prof. L. N. G. Filon FRS (1875–1937), Goldsmid Professor of Applied Mathematics at the University College of London University. In 1934, B. R. Seth was awarded the Ph. D. degree of the University of London for his thesis entitled 'Finite Strain in Elastic Problems'. In between in 1933, B. R. Seth attended some courses at the University of Berlin in Germany.

12.2 Career, Research Contributions

After returning from Europe in 1937, B. R. Seth joined his own alma mater, the Hindu College in Delhi as the Head of the Department of Mathematics. At the same time he was also appointed as a 'Reader' in Mathematics at the Delhi University. He worked in both the Institutions for 12 long years with dedication and distinction. He was able to establish a good tradition of research in Mathematics at the University of Delhi. Many of his students at Hindu College were inspired by him and in later life they became established mathematicians. In the early part of 1949, Prof. B. R. Seth went to Iowa State University in USA for two years. Immediately after he returned to India in 1951, he was appointed as a Professor and the Head of the Department of Mathematics at the newly established Indian Institute of Technology (IIT), Kharagpur in West Bengal. This was India's first such Institute. During his long stay of 16 years at the IIT, Kharagpur, Prof. B. R. Seth successfully built up a strong school of research in Elasticity, Plasticity, Fluid Dynamics, Rheology and Numerical Analysis there. Under his dynamic leadership, this Department of Mathematics gained recognition as a foremost centre of research in these areas nationally as well as internationally.

With great farsightedness and vision, Prof. B. R. Seth established the 'Indian Society of Theoretical and Applied Mechanics (ISTAM)' and nurtured it for 25 years. This was one of India's first interdisciplinary platforms where engineers, applied mathematicians, statisticians and computer scientists met and discussed problems of common interest.

In 1966, Prof. B. R. Seth took charge as the Vice-Chancellor of the newly created Dibrugarh University in Assam. He served there till 1971 and completed the full term. Then he was invited to take over as the Director of the 'Birla Institute of Technology (BITS)' at Mesra, Ranchi in Bihar. He rejuvenated the Institution and worked hard for its 'autonomous status'.

He retired from BITS in 1977 and moved over to Delhi. He was for sometime a visiting 'Fellow' at IIT, Delhi. From 1977, he worked as an 'Emeritus Professor' at the University of Delhi till the last days of his life in 1979.

Throughout these 42 years of his life, Prof. B. R. Seth worked with great dedication to high academic ideals and inspiring leadership. He fought fearlessly for his students and junior colleagues for just causes.

Now a brief discussion is being made of Prof. B. R. Seth's personal research contributions.

Prof. B. R. Seth made notable contributions in three branches of Applied Mathematics: (a) Elasticity, (b) Plasticity and (c) Fluid Mechanics.

A major part of B. R. Seth's thesis was published in the well-known journal titled 'Transactions of the Royal Society'. That reflects the high quality of the research done by him for his Ph. D. degree. In his thesis, he gave second-order treatment of simple torsion, pure flexure and torsion based on quasi-linear stress–strain relation. Many noted researchers such as Elderton, K. Pearson, S. Timoshenko, D. H. Young have contributed to 'Saint-Venant's problem' of torsion and flexure. But they had all considered only the symmetrical case. In 1934, B. R. Seth was successful in giving the first complete solution for a right-angled isosceles triangle. In appreciation of the work done by B. R. Seth, Prof. A. C. Stevenson commented: 'No account of work on Saint-Venant's problem will be complete which did not pay tribute to the work of Seth.'

B. R. Seth also fully solved the problem related to an eccentric hollow shaft. Problems of liquid contained in rotating cylinders with triangular section were solved by him. His first paper on this topic was published in the 'Quarterly Journal of Mathematics, Oxford' in 1934. At the suggestion of the editor, B. R. Seth extended his methods to the region exterior to the cylinder and published a paper on that solution. Subsequently such potential problems both for exterior and interior domains were comprehensively solved by B. R. Seth.

In Fluid Mechanics, two notable contributions of B. R. Seth are as follows. He proved that in a canal of closed section standing transverse waves were not possible. He also suggested modification in Lamb's solution for a canal of circular section, pointing out that the approximate mode assumed was unstable. His other important work was to show that slow viscous solution could be obtained by superposing two effects; (i) one because of irrotational flow for the same boundary and (ii) due to a concentrated force in an infinite liquid, as the drag suffered by the body was equal to the concentrated force.

In addition to his work on Elasticity and Fluid Mechanics, B. R. Seth also collaborated with Prof. F. C. Harris of the University College of London University to carry out experimental work on photo-elastic properties of celluloid.

In 1936, B. R. Seth spent a year at the University of Perugia in Italy. The next year in 1937 he obtained the D. Sc. degree from the University of London for his original contributions in the fields of Elasticity and Fluid Mechanics.

After returning to India B. R. Seth continued with his work on Elasticity. He solved problems of bending, vibration and stability of membranes and plates with triangular boundaries applying finite strain theory to these problems. During the sixties of the last century, he made a very important contribution by determining, among other things, the stage at which the constitutive equations change form. This led to his initiation of the concept of 'transition'. It may be noted that by making a number of ad hoc assumptions like that of incompressibility and a particular form of the yield condition T. Y. Thomas obtained theoretical results on the collapse of thick cylinder. Those results compared well with the experimental results of P. W. Bridgman. But the 'Transition Theory' as developed by B. R. Seth and others readily gives all these results and more general results as well without making any prior assumptions. The paper titled 'Transition theory of elastic plastic deformation, creep and relaxation' [Nature, 195, (1962), 896–897] authored by B. R. Seth needs to be specially mentioned. It may be noted that the classical macroscopic treatment of problems in plasticity, creep and relaxation has to assume semi-empirical yield conditions like those of H. Tresca, von Mises and creep-strain laws like those of F. H. Norton, F. K. G. Odqvist, L. M. Kachanov and others. This is a direct consequence of using linear strain measures, which neglect the nonlinear transition region through which the yield occurs and the fact that creep and relaxation strains are actually never linear. Tresca found out a transient zone and named it as 'mid-zone'. I. Todhunter and K. Pearson supported this finding. In this state the whole of the material participates, and not simply a line or a specified region, as was assumed by the earlier researchers. B. R. Seth in the above mentioned paper pointed out that numerical investigation on flow and deformation theories shows that a continuous approximation through a transition state leads to a satisfactory convergent solution. He in his many other research papers showed that all transition fields like boundary layers, shocks, elastic–plastic deformations, creep and fatigue are subharmonic or super-harmonic in character and that vorticity and spin play a significant role in their growth.

B. R. Seth used asymptotic treatment in finite torsion problem. This provided one of the few examples in which a plastic state which resulted from large deformation could be predicted by elastic theory. This is a noteworthy feature.

To summarize his vast amount of work briefly, it may be noted that his work on finite strain in elastic problems gave a second-order treatment of simple torsion, pure flexure and torsion, based on a quasi-linear stress–strain relation. Prof. B. R. Seth was the first mathematician who successfully gave the first complete solution for a right-angled isosceles triangle. Similarly, he was able to describe the complete solution for an eccentric hollow shaft. He also worked on problems related to bending, vibration and stability of membranes and plates having triangular boundaries applying the finite strain theory.

B. R. Seth has published more than 100 papers in well-known national and international journals. This accounts for his rich contributions to Elasticity, Plasticity, Boundary Layer Theory, Potential Theory and a number of other allied disciplines. His most remarkable work relates to Saint-Venant's Problem, Potential Flows for Exterior and Interior Domains, Distributions of Generalized Singular Points, Systematic Theory of Transition and Non-Linear Measures of Deformation.

Prof. B. R. Seth's important research contributions have been amply referred to in standard books such as 'Mathematical Theory of Elasticity' by I. S. Sokolnikoff [(1956), McGraw-Hill] and 'Elasticity and Plasticity' by J. N. Goodier and P. G. Hodge [(1958), Wiley and Sons]. The extension lectures that Prof. B. R. Seth delivered at the University of Lucknow were published in the form of a monograph titled 'Two Dimensional Boundary Value Problems' by the same University.

Prof. B. R. Seth delivered series of extension lectures on various topics in different Indian Universities. Apart from the lectures he delivered at Lucknow University, which has already been mentioned above, in 1954, he delivered lectures on 'Finite Deformations' at Osmania University, Hyderabad. He again delivered a series of extension lectures on 'Transition Theory of Elastic Plastic Deformations' in 1964 in the same University. In 1965, he delivered a similar series of lectures on 'Some Problems in Fluid Mechanics' at Jadavpur University, Calcutta.

He supervised more than two dozen Ph. D. students, and many of them became famous mathematicians in later life.

12.3   Special Honours and Awards

Apart from his services in Indian Institutions, Prof. B. R. Seth worked as 'Visiting Professor' at three different American Institutions. They are Iowa State University, Ames, Iowa (1949–1950), Mathematics Research Centre, University of Wisconsin, Madison (1961–1962) and Oregon State University, Oregon (1967–1968). He was a very widely travelled mathematician. He visited more than two dozen countries in

Europe, Asia, Australia and North America. He delivered lectures in many famous Universities of the World. He was invited four times by the USSR Academy of Sciences and five times by the Polish Academy of Sciences for delivering special lectures. He attended six International Congresses of Mathematicians and four International Congresses on Theoretical and Applied Mechanics in various foreign cities.

Prof. B. R. Seth was elected as 'Fellow' of the Indian National Science Academy (INSA), Indian Academy of Sciences (Bangalore), Delhi School of Economics and the Institute for Social and Economic Change (Bangalore). He was the first Indian to be elected as a 'Fellow' of the Polish Academy. In 1968 Prof. B. R. Seth was conferred the D. Sc. (Honoris Causa) by IIT, Kharagpur as a recognition of his immense contribution to the all-round development of the Institute. In 1957, he was awarded the 'Euler Medal' by the USSR Academy of Sciences for his outstanding contributions to 'Continuum Mechanics'. In March 1978, he received the coveted 'Dr. B. C. Roy National Award for Eminent Men of Science' for the year 1977. He is the only mathematician in India to have received this award till date.

He died of cardiac illness-related problems on 12 December 1979, in Delhi.

12.4   Important Milestones in the Life of Professor Bhoj Raj Seth

1907: Born on 27th August 1907, in a middle-class Punjabi family at Bhura in Punjab.

1927: Obtained his B. A. (Honours) degree in Mathematics from Hindu College of Delhi University. He stood First with a First class among all the successful candidates from the Delhi University that year. He was awarded the University Gold Medal for his brilliant performance.

1929: Obtained his M. A. degree in Mathematics from Hindu College of Delhi University. He repeated his earlier performance and again stood First with a First class from the Delhi University. He was again awarded the University Gold Medal for his remarkable performance.

1929: On account of his outstanding performance, he was awarded the 'Central Government Scholarship' and went to England, UK, for pursuing higher studies. He joined the Post-Graduate Department of Mathematics at the University College of London University.

1932: He obtained the M. Sc. degree in Mathematics from the University of London.

1933: He attended courses on Mathematics at the University of Berlin, Germany.

1934: He was awarded the Ph. D. degree of the University of London for his thesis titled 'Finite Strain in Elastic Problems'.

1936: He spent a year at the University of Perugia in Italy.

1936: Elected 'Fellow' of the Indian Academy of Sciences (Bangalore).

1937: He obtained the D. Sc. degree from the University of London for his original contributions in Applies Mathematics.

1937: Joined as the Head of the Department of Mathematics in Hindu College, Delhi. Also joined as 'Reader' in the Department of Mathematics, University of Delhi.

1948: Attended the seventh 'International Congress on Theoretical and Applied Mechanics' (ICTAM) at London, UK.

1949–1950: Joined and worked at the Department of Mathematics of the Iowa State University, Ames, Iowa in USA as a 'Visiting Professor'.

1950: Attended 'International Conference of Mathematicians' (ICM) at Harvard, USA.

1951: Joined as Professor and Head of the Department of Mathematics at IIT, Kharagpur in West Bengal, India.

1954: Attended 'International Conference of Mathematicians' (ICM) at Amsterdam, Netherlands.

1955: President of the 'Mathematics Section' of the annual meeting of the Indian Science Congress Association held at Baroda.

1957: Attended 'International Symposium on Boundary Layers' organized by 'International Union of Theoretical and Applied Mechanics' (IUTAM) at Freiberg, Germany (erstwhile East Germany)

1957: The USSR Academy of Sciences awarded him the 'Euler Medal' in recognition of his contributions in Continuum Mechanics.

1958: Attended 'International Symposium on Non-homogeneities in Elasticity and Plasticity' organized by 'International Union of Theoretical and Applied Mechanics' (IUTAM) at Warsaw, Poland.

1958: Attended 'International Congress on Rheology' in West Germany.

1959: Attended 'International Symposium on Elastic–Plastic Shells' organized by 'International Union of Theoretical and Applied Mechanics' (IUTAM) at Delft, Netherlands.

1961: Elected as a 'Fellow' of the Indian National Science Academy (INSA), New Delhi.

1961: Attended 'International Symposium on Secondary Effects' organized by 'International Union of Theoretical and Applied Mechanics' (IUTAM) at Haifa, Israel.

1961–1962: Joined and worked at the 'Mathematics Research Centre', University of Wisconsin, Madison in the USA as 'Visiting Professor'.

1963: Attended 'International Symposium on Continuum Mechanics' organized by 'International Union of Theoretical and Applied Mechanics' (IUTAM) at Tbilisi, Georgia (erstwhile USSR).

1964: Attended the eleventh 'International Congress on Theoretical and Applied Mechanics' (ICTAM) at Montreal, Canada.

1964–1966:President of the 'Indian Society of Theoretical and Applied Mechanics'.

1967–1968: Joined and works at the Department of Mathematics, Oregon State University, Oregon, USA, as a 'Visiting Professor'.

1968: Attended 'International Conference of Mathematicians' (ICM) at Stockholm, Sweden.

1968: The IIT, Kharagpur honoured him by conferring the D. Sc. (Honoris Causa) as recognition of his immense contribution to the Institute.

1969: President of the 'Mathematical Association of India'.

1970: Attended 'International Conference of Mathematicians' (ICM) at Nice, France.

1971: Participated in the Fifth Commonwealth Educational Conference held at Canberra, Australia.

1972: Attended the thirteenth 'International Congress on Theoretical and Applied Mechanics' (ICTAM) at Moscow, USSR.

1972: Attended 'International Congress on Rheology' in France.

1974: Attended 'International Conference of Mathematicians' (ICM) at Vancouver, Canada.

1976: Attended 'International Symposium on Creep' organized by 'International Union of Theoretical and Applied Mechanics' (IUTAM) at Gothenburg, Sweden.

1976: Attended the thirteenth 'International Congress on Theoretical and Applied Mechanics' (ICTAM) at Delft.

1976: Attended 'International Congress on Rheology' in Sweden.

1978: Attended 'International Conference of Mathematicians' (ICM) at Helsinki, Finland.

1978: Honoured by the 'Dr. B. C. Roy National Award for Eminent Men of Science'.

1979: Breathed his last due to heart attack on 12th December 1979, at Delhi.

Chapter 13
Subodh Kumar Chakrabarty (1909–1987)

13.1 Birth, Family, Education

S. K. Chakrabarty was born on 18th July 1909, at Barishal in erstwhile East Bengal (present-day Bangladesh). Having lost his father Sitala Kanta Chakrabarty at an early age of six, he was brought up by his mother Sarala Devi. He completed his early education in East Bengal. He passed his Matriculation Examination in 1925 and the Intermediate Examination in Science (ISc.) in 1927. He graduated with Mathematics (Honours) from Government Brajamohun College at Barishal in 1929. After that he came to Calcutta and got himself enrolled at the famous Department of Applied Mathematics of Calcutta University to pursue Post-Graduate studies. In 1932 he came out with flying colours standing First with a First class not only from the Department of Applied Mathematics, but also securing highest marks among all the successful candidates who had appeared that year for the M. Sc. Examination in different subjects. He was awarded the 'Sir Asutosh Mukherjee Gold Medal' for his superlative performance in the M. Sc. Examination.

13.2 Career, Research Contributions

After completing his Post-Graduate studies S. K. Chakrabarty started teaching at the Mathematics Department of the 'City College' in Calcutta. He taught there and along with that, in 1935 he joined the Department of Applied Mathematics as a part-time teacher and started teaching in the Post-Graduate classes. Then his academic life got a new dimension. In 1940, Dr. H. J. Bhabha came to the Calcutta University to deliver a course of 'Invited Lectures' on cosmic rays. At that time S. K. Chakrabarty was working on some discrete problems of Quantum Mechanics. After listening to Bhabha's lectures and personally interacting with him, he got interested in the 'development of cosmic showers' through the 'cascade processes'. Following the physical ideas put forth by Bhabha and Heitler and also by Carlson and Openheimer,

© The Author(s), under exclusive license to Springer Nature Singapore Pte Ltd. 2022   111
P. Mukherji, *Notable Modern Indian Mathematicians and Statisticians*,
https://doi.org/10.1007/978-981-19-6132-8_13

Bhabha and Chakrabarty put forward an elegant and rigorous analysis for the accurate estimation for 'shower particles and quanta' at different depths, produced by an electron and a photon, in passing through solid materials or atmosphere, making a more rigorous treatment of the problem, from the standpoints of both physical assumptions and mathematical treatments. In a series of investigations, Bhabha and Chakrabarty studied the effect of 'collision loss' and derived solutions for the 'cascade equations', which may be used for estimating the energy distribution of the 'shower particles and photons' of all energies, including those near and even below the 'critical energy'. Bhabha was highly impressed by the mathematical ingenuity of S. K. Chakrabarty and extended an invitation to him to visit Bangalore, where Bhabha had set up his own research unit. S. K. Chakrabarty accepted the invitation and visited Bangalore during February—June, 1941. Their collaborative research produced the results mentioned above.

S. K. Chakrabarty soon won many accolades for his brilliant research work. He was the recipient of the prestigious Premchand Roychand Scholarship (PRS) of the Calcutta University. In 1943 he obtained the D. Sc. degree of the Calcutta University for his original contributions in Mathematical Physics. He received the 'Mouat Medal' from the Calcutta University the same year. In 1944, the 'Royal Asiatic Society' awarded him the 'Elliot Prize' in appreciation of his innovative research work.

In 1945, Dr. S. K. Chakrabarty left Calcutta and joined as Director of Colaba and Alibag Observatories in Bombay. In the meantime, Dr. H. J. Bhabha had joined the Tata Institute of Fundamental Research (TIFR) as the Director. This was the time when Chakrabarty and Bhabha again collaborated and developed the famous theory of 'Cascade showers in cosmic radiation'. In subsequent years, they jointly made significant contributions in the Dynamo Theory of S_q variations, relations between geomagnetic storms and emissions of solar corpuscles, earthquake source mechanisms from the analysis of seismograms. The interaction between these two scientists remained firmly entrenched on fruitful scientific exchanges and ended only with the untimely demise of Dr. Bhabha.

In 1948, Dr. S. K. Chakrabarty went to USA and worked for sometime at the California Institute of Technology (CALTECH), Pasadena as a 'Visiting Research Fellow'. There he had the opportunity of working with a group of eminent seismologists under the leadership of Prof. B. Gutenberg in collaboration with C. F. Richter. He did a substantial amount of theoretical work by analysing the records of the 'Walker Pass earthquakes' of 15th March 1946. Around that time, Dr. Chakrabarty also started his research on the response characteristics of electromagnetic seismographs and their precise estimation through experiments carried out in collaboration with Dr. Benioff, at the Seismological Laboratory, Pasadena.

A summary of some of these works in which Dr. S. K. Chakrabarty was involved was incorporated in the UNESCO publication titled '*Manual of Seismological Observatory Practice*'. This report was compiled and prepared by the International Committee for the 'Standardization of Seismographs and Seismograms'. This 'Committee' had been formed as a part of a UNESCO resolution.

On his return to India, Dr. Chakrabarty after working with the 'Meteorological Department' of the Government of India for sometime decided to resign and return to Calcutta. He finally joined the Bengal Engineering College (B. E. College) [present-day ISET], Shibpur, Howrah, a suburb near Calcutta in 1949. He took charge as Professor and Head of the Department of Mathematics and served there till 1963. While at B. E. College, Professor Chakrabarty took a very proactive role in different academic and administrative matters. Under the sponsorship of the 'Council of Scientific and Industrial Research' (CSIR), Prof. S. K. Chakrabarty established a 'Seismographic station' at the premises of the B. E. College and conducted study of microseism and sea waves and their correlation with cyclonic disturbances. In a sense Prof. S. K. Chakrabarty was one of the pioneer researchers on 'Theoretical Seismology' in India. He tried to establish a school of research on 'Theoretical Seismology and Geophysics' in Bengal. Apart from Cosmology, Seismology and Atmospheric Sciences were his areas of interest. Personally, Prof. S. K. Chakrabarty published 39 research papers in reputed national and international journals. It may be noted that 3 research papers were written by him in German and published in 'Verlag Von Julius, Springer (VVJS)' a famous German scientific Journal. The three papers are listed below:

- *'Stark-Effekt des rotationsspektrums and electrischesuzeptibilitatbeihoher temperature'* [VVJS, Berlin Sonderbdruck, 102, Band 1 and 2, Heft (1936), 102–111].
- *'Das eigenwert-problem, eineszweiartomigenmolekuls and die berechung der dissoziationsenergie'* [VVJS, Berlin Sonderbdruck, Zeitschrift fur Physic (ZFP), 109, Band 1, 2 Heft (1937), 25–38].
- *'Notizuber den stark effect der rotationsspektren'* [VVJS, Berlin Sonderbdruck, Zeitschrift fur Physic (ZFP), 110 Band 11, 12 Heft (1938), 688–691].

Apparently these are on problems related to Quantum Mechanics. His collaborative research paper with H. J. Bhabha titled *'The cascade theory with collision loss'* [Proc. Royal Soc., London, A, 181, (1943), 267–303] is a commendable publication.

The following 2 research papers mentioned below have been internationally acclaimed. They are:

- *'Cascade showers under thin layers of materials'* [Nature, London, 158, (1946), 166].
- *'Sudden commencements in geomagnetic field variations'* [Nature, London, 167, (1951), 31].

Prof. S. K. Chakrabarty was the author of the 3 books listed below:

- *'Teaching of Mathematics in Secondary Schools'* [West Bengal Board of Secondary Education (1974)].
- *'Elements of Discrete Mathematics (with applications to Computer Science)'* [Allied Publishers Ltd., (1976)].
- *'A teacher's commentary on elements of Discrete Mathematics'* [Allied Publishers Ltd., (1978)].

In 1963, Prof. S. K. Chakrabarty joined his alma mater, the Department of Applied Mathematics, University of Calcutta as the 'Rashbehary Ghosh Professor' and the Head of the Department. He served there till his retirement in 1974.

During his tenure at the University College of Science, Calcutta University, he tried very hard to encourage research in the areas of 'Theoretical Seismology' and 'Mathematical Geophysics'. With his vast knowledge in these disciplines and the past experience of his collaborative research with such stalwart geophysicists as Professors B. Gutenberg, C. F. Richter, V. H. Benioff, J. A. Fleming among others made Prof. S. K. Chakarabarty an iconic expert in the field. Prof. Chakrabarty also tried to initiate serious research on topics like Discrete Mathematics with applications to Computer Science, Automata Theory, Artificial Intelligence, etc. He had arranged for the installation of an IBM computer at the premises of the University College of Science, but the project had to be abandoned due to political interference. This turned out to be a great technological setback not only for the University of Calcutta, but for the entire state of West Bengal.

13.3　Special Honours

Prof. S. K. Chakrabarty was elected a 'Fellow' of the National Academy of Sciences (present-day Indian National Science Academy) in 1949. He was Chairman of the Cosmic Ray Research Committee of the Atomic Energy Commission, Government of India for many years. He was elected the President of the Mathematics Section of the annual Indian Science Congress Association session held in 1954 at Hyderabad. Prof. Chakrabarty was the President of the Calcutta Mathematical Society from 1970 to 1972. He was the Founder Chairman of the 'Advanced Study Centre of Mathematics' in the Department of Applied Mathematics, University College of Science, Calcutta University.

Prof. S. K. Chakrabarty was an eminent scholar, a distinguished scientist and a pioneer researcher in the field of 'Mathematical Geophysics' and 'Atmospheric Sciences' in India. He passed away in Calcutta in the early hours of the 14th November 1987, after a brief illness.

13.4　Important Milestones in the Life of Professor Subodh Kumar Chakrabarty

1909: Born on 18th July 1909, at Barishal in erstwhile East Bengal.
1929: Graduated with Mathematics (Honours) from Barishal Brajamohun College.
1932: Obtained the M. Sc. in Applied Mathematics standing First with a First class from the University of Calcutta. Awarded the 'Sir Asutosh Mukherjee Gold Medal'

by the University of Calcutta for securing highest marks among all successful students in all subjects.

1943: Obtained the D. Sc. degree from the University of Calcutta for original contributions in Mathematical Physics.

1945–1948: Worked as Director of Colaba and Alibag Observatories, Bombay.

1948: Proceeded to California Institute of Technology (CALTECH), Pasadena, USA as Visiting Research Fellow.

1949: Joined the Bengal Engineering College (BE College), Shibpur, Howrah [present-day ISET] as Professor and Head of the Department of Mathematics.

1949: Elected 'Fellow' of the National Institute of Sciences [present-day 'Indian National Science Academy' (INSA)].

1954: Elected 'President' of the Mathematics Section of the Indian Science Congress held at Hyderabad.

1963: Joined the University College of Science, Calcutta University as the 'Rashbehary Ghosh Professor' and Head of the Department of Applied Mathematics.

1970–1972: Elected 'President' of the Calcutta Mathematical Society.

1974: Retired from the University of Calcutta.

1987: Passed away in Calcutta on the 14th November 1987, after a brief illness.

Chapter 14
Manindra Chandra Chaki (1913–2007)

14.1 Birth, Education

Manindra Chandra Chaki (M. C. Chaki) was born on the 1st July 1913, at the village Deuli of Bagura district in erstwhile East Bengal (present-day Bangladesh). His father Keshab Chandra Chaki was a landlord and mother Kunjakamini Debi a pious and kind-hearted housewife. M. C. Chaki had his school education in East Bengal. It may be noted that in his early school days Mathematics did not interest him much. But under the inspiration of his private tutor Sri Durgadas Banerjee, he developed a liking for the subject. In 1930, M. C. Chaki completed his Matriculation Examination from Gaibandha High School situated in the Rangpur district, with a first division. He scored more than 75% marks in the aggregate. After that he moved over to Calcutta and got himself admitted to Bangabasi College, a private College in the city. In 1932, M. C. Chaki passed the Intermediate Examination in Science (I. Sc.) from this College with a first division. After that he took admission in the Rajshahi Government College in Rajshahi, East Bengal. Initially, he enrolled for the B. A. class with Honours in English. Then Prof. B. M. Sen (1888–1978), himself a reputed mathematician, was the Principal of that College. He advised M. C. Chaki to join the Mathematics Honours course. Under his advice, Chaki withdrew from the English class and joined the B. A. (Honours) class in Mathematics with Sanskrit as the subsidiary subject. In 1934 he graduated with second class in the B. A. Mathematics (Honours) Examination. Thereafter, he joined the Department of Pure Mathematics in the University of Calcutta to pursue Post-Graduate studies. In 1936, he passed the M. A. Examination in Pure Mathematics standing Second with a First class from the University of Calcutta.

After that, at the suggestion of his uncle, M. C. Chaki attended the law classes of Calcutta University for a brief period. But he was not at all interested in the subject, and finally he gave it up. Thereafter, he taught as part-time Lecturers in various private colleges for some time. Finally, he rejoined his alma mater, the Department of Pure Mathematics in the University of Calcutta as a part-time Lecturer in 1950.

© The Author(s), under exclusive license to Springer Nature Singapore Pte Ltd. 2022   117
P. Mukherji, *Notable Modern Indian Mathematicians and Statisticians*,
https://doi.org/10.1007/978-981-19-6132-8_14

He also started doing research in Mathematics under the guidance of the famous geometer and the Hardinge Professor of Pure Mathematics Prof. R. N. Sen.

14.2   Career, Research Contributions

As regards his teaching career, after completing his M. A. in Pure Mathematics, M. C. Chaki first started working from 1939, as a Lecturer in Mathematics and English at the Azizul Haque College in Bagura, East Bengal. He served there till 1945. In 1945, he came to Calcutta and joined the Bangabasi College as a Lecturer in Mathematics and continued to teach there till 1952. He also worked as a part-time Lecturer in Mathematics in his own alma mater, the Department of Pure Mathematics of the University of Calcutta during 1950–1952. Finally in 1952, he was recruited in the same Department as a full time 'Lecturer' in Pure Mathematics. In 1960 he became a 'Reader' in Pure Mathematics and served in that capacity during 1960–1972. In 1972, he was selected as the 'Sir Asutosh Birth Centenary Professor of Higher Mathematics' (formerly known as 'Hardinge Professor of Higher Mathematics') in the Department of Pure Mathematics, University of Calcutta. He adorned the Chair Professor's post till his retirement in 1978. As a teacher he was quite successful and popular in the student community.

Going back to M. C. Chaki's research contributions, it may be noted that he obtained his D. Phil degree in Pure Mathematics for his thesis entitled 'On some problems in Riemannian Geometry', in 1956 from the University of Calcutta. One of the examiners of the thesis, Prof. L. P. Eisenhart of Princeton University, USA. highly praised the work. Dr. M. C. Chaki was a prolific researcher in various branches of Geometry. He made notable contributions in the fields of Riemannian Geometry, Classical Differential Geometry, Modern Differential Geometry, Theoretical Physics, General Relativity and Cosmology and History of Mathematics. More than 20 students obtained their Ph. D. degrees under his supervision. He personally published more than 60 research papers, which were published in reputed national and international journals. Now some of his more important papers published during the twentieth century are being discussed. In the paper titled '*On a non-symmetric Harmonic Space*' [Bulletin of the Calcutta Mathematical Society, 44, (1952), 37–40], Prof. Chaki constructed an example of a simple harmonic space having dimension n, $n > 4$, which is neither flat nor symmetric in the sense of Cartan, but it is recurrent. This led to the nullification of a conjecture made by A. Lichnero-wicz (1915–1998), a renowned French geometrician. According to his conjecture, the dimension of a harmonic space must be less than or equal to 4, failing which it will be a Cartan-symmetric space. This paper of Prof. Chaki has been internationally acclaimed and has been referred to in the book titled 'Harmonic Spaces' [authored by H. S. Ruse and A. G. Walker; Roma, Edizioni cremonose (1961)]. Subsequently a lot of research was done in this area by many researchers both in India and abroad. In 1963, in the paper titled '*On conformally symmetric spaces*' [(with Bandana Gupta) Indian Journal of Mathematics, 5 (2), (1963), 113–122], Prof. Chaki in collaboration with his first research scholar Miss Bandana Gupta introduced a new type of Riemannian Space called conformally symmetric space. Two important results have been proved in this paper. They are as follows:

- A conformally symmetric space is of constant scalar curvature if and only if the first covariant derivative of Ricci tensor is a symmetric tensor.
- Every Einstein conformally symmetric space is symmetric in the sense of Cartan.

This research paper made notable impact on researchers in Differential Geometry. Polish and Japanese researchers have extensively worked in this Riemannian Space. The paper was found to be useful in the General Theory of Relativity. Some research workers in USA have studied conformally symmetric spaces with semi-Riemannian metric or indefinite metric. It may be noted that in Mathematical Physics, the conformal symmetry of space–time is expressed by an extension of the Poincare group. The extension includes special conformal transformations and dilations.

Another interesting fact is that, when Prof. M. C. Chaki initiated his studies on Riemannian Geometry he used Eisenhart's index method for representing tensors. But from early eighties of the twentieth century, his interest shifted to global Differential Geometry of manifolds. From then on he wrote his research papers using modern index-free notations and advised his students to do the same. In 1987, Prof. Chaki introduced the notion of Pseudo-Symmetric Manifolds, in his now famous research paper titled '*On Pseudo-Symmetric Manifolds*' [Analele Stn. AL. I. Cuzo. Din Lasi, Tome XXXIII (1), (1987), 53–58]. He used this new concept and obtained six important theorems. He dedicated this particular research paper to the famous Japanese geometrician Kentaro Yano (1912–1993) on the occasion of the latter's 74th birthday. In modern mathematical literature this manifold has been named as 'Chaki Manifold' by Prof. M. Toomanian [Mathematical Reviews, USA, September, (1993), 4970.]. Symbolically, an n-dimensional 'Chaki manifold' is now denoted by Chaki $(PS)_n$.

More than 20 students obtained their Ph. D. degrees under the supervision of Prof. M. C. Chaki. Among them, Dr. Bandana Barua (nee Gupta), Dr. Dipak Kumar Ghosh, Prof. U. C. De and Prof. M. Tarafdar served as Faculty Members in the Department of Pure Mathematics, University of Calcutta. Among Prof. Chaki's other students, Dr. A. N. Raychowdhury served as a Professor at NIT, Durgapur, Prof. Asoke K. Roy was a Professor at Jadavpur University, Dr. K. K. Sharma was a Professor at Tripura University, and Dr. A. Konar served as a Professor at Kalyani University. Dr. A. K. Bag and Dr. Pradip Kumar Majumdar did their Ph. D. s in 'History of Mathematics' under the guidance of Prof. Chaki. Dr. Bag was the Head of the 'History of Science Division' of the Indian National Science Academy at New Delhi. He was also the Chief Editor of 'Indian Journal of History of Science'. Dr. Majumdar was Professor of Indian Astronomy at the 'School of Vedic Studies' at Rabindra Bharati University, Calcutta.

14.3 Special Honours

In 1951 Prof. Chaki was elected as a Fellow of the Royal Astronomical Society, London. Since 1964 he served as a 'reviewer' of 'Mathematical Reviews', USA. In the 63rd Session of the Indian Science Congress, held at Visakhapatnam in 1976, Prof. Chaki was the President of the Mathematics Section. In 1981, he served as the President of the Calcutta Mathematical Society. He was a Member of the editorial board of the journal 'Tensor', Japan. He was an Honorary Fellow of the Asiatic Society, Calcutta. In 1993, Prof. M. C. Chaki was honoured as 'Teacher of Eminence' by the University of Calcutta.

This renowned geometer and a much admired teacher of Mathematics, Prof. M. C. Chaki breathed his last on 21st July 2007, in Calcutta.

14.4 Important Milestones in the Life of Professor Manindra Chandra Chaki

1913: Born on 1st July 1913, at a small village in the district of Bagura in erstwhile East Bengal (present-day Bangladesh).

1936: Passed M. A. in Pure Mathematics standing Second with a First class from the University of Calcutta.

1952: Joined the Department of Pure Mathematics, University of Calcutta as a Lecturer.

1951: Elected Fellow of the Royal Astronomical Society.

1956: Obtained D. Phil in Pure Mathematics from the University of Calcutta.

1960: Became a 'Reader' in Pure Mathematics at the University of Calcutta.

1972: Selected as 'Sir Asutosh Birth Centenary Professor of Higher Mathematics' at the Department of Pure Mathematics, University of Calcutta.

1976: Elected President of the Mathematics Section of the 63rd Indian Science Congress held at Visakhapatnam in 1976.

1978: Retired from the University of Calcutta.

1981: Served as President of the Calcutta Mathematical Society.

1993: Honoured as 'Teacher of Eminence' by the University of Calcutta.

2007: Passed away in Calcutta on 21st July 2007.

Chapter 15
Calyampudi Radhakrishna Rao FRS (Born: 10 September 1920)

15.1 Birth, Childhood, Education

Calyampudi Radhakrishna Rao (C. R. Rao) was born on 10 September 1920 in a small town named Huvvinna Hadagalli in erstwhile Madras Presidency of British-ruled India (present-day state of Karnataka). As C. R. Rao was the eighth child of his parents C. D. Naidu and A. Lakshmikanthamma, he was named 'Radhakrishna' after the Hindu deity Sri Krishna. C. D. Naidu was an inspector in the 'Criminal Investigation Department'. He recognized C. R. Rao's special aptitude for Mathematics quite early and encouraged his son to study Mathematics seriously and pursue research in that discipline as a career option. In later life, C. D. Naidu always had great expectations from his brilliant son and confessed that he had high hopes about C. R. Rao's achievements and often said C. R. Rao was his 'pride, hope and joy.' His mother instilled in him the discipline and work-related ethics. C. R. Rao dedicated his book titled 'Statistics and Truth: Putting Chance to Work' [Ramanujan Memorial Lectures, Reprint Edition, Council of Scientific and Industrial Research, New Delhi, (1989)] to his mother and acknowledging her role and influence in his life wrote: 'For instilling in me the quest for knowledge, I owe to my mother, A. Lakshmikantamma, who, in my younger days, woke me up every day at four in the morning and lit the oil lamp for me to study in the quiet hours of the morning when the mind is fresh.' His mother was a strict disciplinarian and punished the child C. R. Rao when he missed the school without valid reasons. Eventually C. R. Rao became more mature and disciplined and did very well in all the School Examinations. During his school career, in 1935, he won the 'Chandrasekhara Scholarship' in Physics. This was supposed to be the most coveted prize throughout his school career. Because of the transferable nature of his father's job, C. R. Rao had to study in various schools. Finally, after retirement in 1931, his father settled down in Visakhapatnam, in the present-day coastal Andhra region. After successful completion of his school education C. R. Rao joined the 'Mrs. A. V.N.College' there and passed his Intermediate Examination in Science (I. Sc.) with Physics, Chemistry and Mathematics. He did very well in all the subjects,

© The Author(s), under exclusive license to Springer Nature Singapore Pte Ltd. 2022
P. Mukherji, *Notable Modern Indian Mathematicians and Statisticians*,
https://doi.org/10.1007/978-981-19-6132-8_15

but his superlative performance in Mathematics reflected his unusual prowess in the subject. It may be noted that another outstanding Indian scientist, the Nobel Laureate Sir C. V. Raman, had also completed his I. Sc. Examination from the same college.

At his father's suggestion, C. R. Rao continued his studies in Mathematics, and in 1940 he earned his M. A. degree in Mathematics from the Andhra University, Waltair (Visakhapatnam). In the said Examination, he stood First with a First class.

Because of the ongoing Second World War in Europe, jobs in Mathematics were scarce. So C. R. Rao decided to go to Calcutta for appearing in an interview in the Department of Survey. While travelling, he casually met a young man, who had recently enrolled himself in a training programme in Statistics at the Indian Statistical Institute (ISI), Calcutta. For gaining more information related to the course, C. R. Rao went to ISI and met the research personnel there. He thus learnt about the ongoing projects there and became convinced that the above-mentioned training programme will be extremely beneficial for him. He felt that it would enhance his job prospects and also open up fresh avenues for a research career. In an interview published by the 'Indian Academy of Science' and titled 'Wise Decisions under uncertainty: an interview with C. R. Rao' [B. V. R. Bhat; Resonance: Journal of Science Educ., 18, (2013), 1127–1132] Prof. Rao stated: 'I went back to Visakhapatnam and told my mother that the only alternative for me was to get admitted to ISI for training in Statistics, and it would cost me Rs. 30/- a month to stay in Calcutta. She said that she would raise the money somehow and that I should go to Calcutta to join ISI. I travelled to Calcutta with Rs. 30/- in my pocket and joined ISI on 1 January, 1941.' That was a turning point in his life.

ISI under the dynamic leadership of the great visionary and legendary statistician Prof. P. C. Mahalanobis offered a training programme in Statistics, which was quite unique in India of those times. This programme attracted students from all over the country and from various disciplines. Government officials who wanted to apply statistical techniques for data analysis and modelling also attended this ISI-sponsored course. Though C. R. Rao found the programme 'somewhat disorganised' as far as teaching was concerned, but he felt fortunate as he had the privilege of coming in contact with the outstanding research Faculty at ISI of those times. There were such stalwarts as R. C. Bose, S. N. Roy, K. R. Nair and some others. In the meantime, Calcutta University, relentlessly pursued by Prof. P. C. Mahalanobis finally started India's first Post-Graduate programme in Statistics in 1941 and selected Prof. P. C. Mahalanobis to be the Honorary Head of the newly established Department. Recognizing the great potential of C. R. Rao, Prof. Mahalanobis urged him to get enrolled for the newly launched M. A. course in Statistics at the University of Calcutta. So C. R. Rao joined the first batch of students at the University. In 1943, he obtained his M. A. degree in Statistics from the University of Calcutta, standing First with a First class. He secured 87% of the total marks, a record unbroken till date. He was awarded the University Gold Medal for his outstanding performance in the M. A. Examination. C. R. Rao created a record as he was among the first 5 people to earn a Post-Graduate degree in Statistics in India from any Indian University. The dissertation for the master's degree which was submitted by the young 23 year old C. R. Rao was an extraordinary piece of scholastic work. In an interview with Anil

Kumar Bera [A. K. Bera; 'The ET interview: Professor C. R. Rao', Economic Theory, 19, (2003), 331–400], Professor Rao stated: 'Looking through my thesis, forwarded by Prof. P. C. Mahalanobis on June 18, 1943, to the Controller of Examinations of Calcutta University, I find that I must have been working hard during the period 1941–1943. The thesis was in three parts, the first with 119 pages on design of experiments, the second with 28 pages on multivariate tests, and the third with 42 pages on bivariate distributions.' According to one of the examiners of the dissertation, 'the work was almost equivalent to a Ph. D. degree.' His write-up had original contributions, which included a solution to a characterization problem formulated by the renowned Norwegian economist Ragnar Frisch. In 1949, this solution was published as 'Note on a problem of Ragnar Frisch' [Econometrica, 15, (1949), 245–249].

15.2 Career, Research Contributions

After earning his M. A. degree in Statistics, from the University of Calcutta, C. R. Rao joined the ISI as a 'Technical Apprentice'. He continued with his research work and taught as a part-time teacher at the Statistics Department of the Calcutta University. The early success story in research of C. R. Rao relates to his collaborative work with two stalwart mathematician-cum-statisticians at ISI, namely R. C. Bose and S. N. Roy. During the period 1944–1946, C. R. Rao published a number of research papers in different areas of Statistics. He used the theory of Combinatorics in experimental design, linear estimation and Multivariate Analysis. He was still working mainly on mathematical Statistics. Around that time Prof. P. C. Mahalanobis asked him to be the leader of a project analysing multivariate anthropometric data collected in Uttar Pradesh at the time of 1941 census in the said state. This experience deeply influenced C. R. Rao and inspired him to work seriously on the interplay between applications and theory. Throughout his long career, this remained his central doctrine. While C. R. Rao was working on that project, J. C. Trevor, an anthropologist at the Duckworth Laboratory at Cambridge requested Prof. Mahalanobis to depute someone from ISI to help him in his work related to the analysis of ancient skeletal remains unearthed by a British expedition in Sudan. Prof. Mahalanobis, the life-long mentor of C. R. Rao, recommended his favourite protégé for the assignment. So in August 1946 C. R. Rao sailed for Cambridge, England. He started working at the Anthropological Museum and simultaneously joined the King's College of Cambridge University to carry out research under the guidance of Prof. R. A. Fisher FRS (1890–1962), one of the greatest statisticians of all times. Prof. Fisher was then the 'Balfour Chair Professor of Genetics' at the University of Cambridge.

Even at that young age of 26, C. R. Rao was a man who knew how to make the best use of his opportunities. He learnt genetics from Fisher, took a course in stochastic processes from the famous British statistician M. S. Barlett FRS (1910–2002) and got involved in an economics reading group studying the path-breaking book titled 'Theory of Games and Economic Behaviour' [authored by John Von Neumann and Oskar Morgenstern; Princeton University Press (1944)]. He even managed to attend the Cambridge Debating Society events and thus got the exposure to some of the leading scientists, philosophers and politicians of the times including the world-renowned mathematician Bertrand A. W. Russell FRS (1872–1970). In those days, C. R. Rao had full-time work at the Anthropological Museum. This involved the statistical analysis of measurements made on skeletons excavated in Jebel Moya in far-away Sudan. On top of that, as instructed by Prof. Fisher, Rao started conducting some experiments with mice for mapping of chromosomes. Based on the work carried out at the museum, C. R. Rao submitted a thesis titled 'Statistical Problems of Biological Classification' to the University of Cambridge. He obtained his Ph. D. degree in Statistics from the University in 1948. Years later in 1965, C. R. Rao earned the degree Sc. D. from the University of Cambridge, UK, for his overall contributions to theory and applications in Statistics.

It would be relevant to discuss some of the major contributions that C. R. Rao made during the period 1945–1965. Just before going to England, in 1945, he had published the paper titled *'Information and accuracy attainable in the estimation of statistical parameters'* [Bulletin of the Calcutta Mathematical Society, 37 (3), (1945), 81–91]. This piece of research established his credibility as one of the legends of modern Statistics. Written in a lucid fashion with elegant proofs, the article gave two foundational results in statistical inference. Later it also paved the way for the growth into the field of Information Geometry. In his two papers titled *'On the mathematical foundation of theoretical statistics'* [Philosophical Transactions of the Royal Society, A, 222, (1921), 309–368] and *'Theory of statistical estimation'* [Mathematical Proceedings of the Cambridge Philosophical Society, 22, (1925), 700–725], Prof. Fisher established the concept of 'Fisher Transformation'. In the paper published in 1945 mentioned above, Rao was able to show that any unbiased estimator of a parameter has a variance that is bounded below by the reciprocal of 'Fisher Transformation'. He obtained the result by using a simple application of the Cauchy-Schwartz inequality. This became one of the most celebrated theorems in Statistics. However, the Swedish mathematician Harold Cramer had established the information inequality and this appeared in his renowned book titled 'Mathematical methods of Statistics' [Princeton University Press, New Jersey, (1946)]. C. R. Rao was unaware of this discovery. Later the inequality came to be known as 'Cramer-Rao Lower Bound'. This lower bound is of great importance in the field of signal processing and has been extensively used in many areas of science and engineering. The second fundamental result which was an offshoot of Rao's 1945 research paper relates to the improvement of the efficiency of an estimator. In 1947, in his paper titled *'Conditional expectation and unbiased sequential expectation'* [Annals of Mathematical Statistics, 18, (1947), 105–110], David Blackwell had proved the same result independently. This result is known in modern statistical literature as 'Rao-Blackwell Theorem'.

C. R. Rao's 1945 paper was also one of the earliest to apply differential geometric approaches to probabilistic models. This created the framework for the new field of information theory. He viewed the parametric family as Riemannian manifold with the Fisher information matrix as the associated Riemannian metric tensor. The geodesic distance induced by this Riemannian metric was proposed by Rao as a measure of dissimilarity between two probability distributions. This Riemannian geodesic metric distance is known in statistical literature as 'Fisher-Rao distance'.

Another breakthrough concept by C. R. Rao involves his now famous 'The Score Test'. In his two research papers titled 'Large sample tests of statistical hypotheses concerning several parameters with applications to problems of estimation' [Proceedings of the Cambridge Philosophical Society, 44, (1948), 50–57] and 'Method of scoring linkage data giving simultaneous segregation of three factors' [Heredity, 4, (1950), 37–59], Rao unveiled a new technique for statisticians. Though it took several years for the 'test' to be a part of the mainstream statistical literature, but at present times C. R. Rao's 'Score Test' is part of every statistician's toolkit.

Another remarkable piece of research done by C. R. Rao led to the concept of generalized inverses. In 1954, Prof. C. R. Rao was requested to analyse the long-term effects of radiation on atomic explosions related victims in Hiroshima and Nagasaki. The statistical analysis carried out by him led to the finding of an inverse of $X'X$, where X represented the model matrix in the usual linear model. But $X'X$ was singular. So the inverse was undefined. In his paper titled '*Analysis of dispersion for multiply classified data with unequal number of cells*' [Sankhyā, Ser. A, 15, (1955), 253–280], Rao introduced the concept of 'pseudoinverse'. That same year, R. A. Penrose published a paper on generalized inverses [Ref: Proceedings of the Cambridge Philosophical Society, 51, (1955), 406–413]. After going through Penrose's paper and doing further investigations Prof. Rao was able to discover the key condition for a generalized inverse. In his famous research paper titled '*A note on generalized inverse of a matrix with applications to problems in mathematical statistics*' [Journal of the Royal Statistical Society, Ser. B, 24, (1962), 152–158] the calculus of g-inverses (the abbreviation of generalized inverses as coined by Prof. Rao) and the unified theory of linear estimation was introduced by Prof. Rao in details. With the help of his colleague Sujit Kumar Mitra at ISI, Calcutta, the subject of g-inverse was fully developed by Prof. C. R. Rao, and they together wrote a monograph titled 'Generalized Inverse of Matrices and its Applications' [Wiley, New York, (1971)].

In his almost eight decades long career Prof. C. R. Rao has made phenomenal contributions in many areas of theoretical and applied Statistics. His researches in different other disciplines such as Economics, Electrical engineering, Anthropology and Genetics have influenced generations of researchers and opened up new avenues for them. Briefly it may be stated that Prof. Rao by developing estimation theory in small samples extended the scope of statistical techniques in practice. His name is associated with many results in this area such as the Cramer-Rao lower bound, Rao-Blackwell theorem, Fisher-Rao theorem, Rao's second-order efficiency and the Geary-Rao theorem. In connection with Coding Theory and in experimental design, Rao's research on orthogonal arrays deserves special mention. He has also made pioneering contributions to the development of multivariate statistical analysis. Rao's score statistic and the Neyman-Rao statistic are very useful asymptotic tests of hypotheses. As already discussed Prof. Rao was a pioneer in using differential geometric techniques in problems related to statistical inference, based on Rao's distance function. He has published more than 30 books in the disciplines of

Mathematics, Statistics and Econometrics. His research papers number more than 350 in Mathematics, Statistics and Probability with special emphasis on statistical estimation theory, multivariate analysis, characterization problems, combinatorics and design of experiments, differential geometric methods in problems of statistics and mathematical genetics, generalized inverses of matrices and matrix methods for linear models. His first book titled 'Advanced Statistical Methods in Biometric Research' [Wiley, New York, (1952)] written based on his work at Cambridge is considered to be a primer on multivariate methods and its applications.

Now going back to his career, after completing his work in Cambridge, Dr. C. R. Rao returned to ISI, Calcutta, in August 1948. Prof. P. C. Mahalanobis recruited him as a Professor and also made him the Head of the ongoing 'Research and Training School'. This shows the amount of confidence Prof. Mahalanobis had on his young 29 year old protégé. Prof. C. R. Rao did full justice to that. Under his enthusiastic leadership, the 'Training School' was revamped and teaching and research programmes became more organized. The progress was spectacular. It soon earned country-wide reputation as an excellent centre for advanced study, research and consultation in Statistics, Probability and allied disciplines. ISI became the destination of brilliant students and practitioners of Statistics from all corners of India. Eagerness to learn statistical theory and its applications in different areas were their sole objective. Prof. C. R. Rao boldly and with all dedication took up the challenges involved and supervised many aspiring research scholars. He soon acquired the reputation of an excellent teacher and a very inspiring research guide. In 1950, Prof. Rao started supervising his first Ph.D. student D. Basu. Basu completed his Ph. D. in 1953 and later made fundamental contributions to statistical inference. Over the years about 50 students earned their Ph. Ds. under the guidance of Prof. C. R. Rao. He had the unique capability of supervising different students in completely different areas of Statistics at any given time. He supervised students on various disciplines connected with Probability, Statistics, Econometrics and Mathematics. The topics comprised survey sampling, multivariate analysis, quality control, characterization, experimental design, directional data, combinatorics and graph theory. He also guided students on areas such as statistical genetics, probability theory and stochastic processes and game theory. In a brief write-up some of his world-renowned students have been discussed in Chap. 18.

After India gained her independence, Prof. Rao re-structured the 'Research and Training Programme' at ISI to address the needs of the government and industry. Thus new courses on statistical quality control and industrial statistics were included in the training programme. In 1953, ISI, Calcutta, set up a new division of 'Statistical Quality Control' to provide consultancy services to industries. ISI also set up branches for this work in other parts of the country. Prof. C. R. Rao served the 'Research and Training School' as Professor and Head from 1949 to 1963. From 1963 to 1972, he served as the Director of the 'Training School'. He was also involved with the International Statistical Educational Centre (ISEC), a joint venture of ISI and the International Statistical Institute with support from UNESCO and the Government of India. Prof. Rao directed their training programmes too. ISEC has continued to be an important training centre offering courses in theoretical and applied statistics, to participants from countries in the Middle East, Far East, Southeast Asia and Africa. After the sad demise of Prof. P. C. Mahalanobis in 1972, Prof. C. R. Rao took over as Secretary of the Indian Statistical Institute and served in that capacity from 1972 to 1976. During that period, Prof. C. R. Rao took some special initiatives and a new constitution was framed for ISI, which helped the institute to evolve and grow as a premier research institution. From 1976 to 1984 he continued to work at ISI as a 'Jawaharlal Nehru Professor'. During the period 1987 to 1992, Prof. C. R. Rao held the National Professorship in India. However, in 1979 he took leave from ISI and went to USA. He finally and technically retired from ISI in 1984.

Though he left India in 1979, still for the completeness from the historical point of view, it may be noted that in 1979, he joined as a University Professor at the University of Pittsburgh. In 1988 he joined the Pennsylvania State University as the 'Eberly Family Chair' Professor in Statistics. He still continues to work in that capacity in the Department of Statistics at the Pennsylvania State University. He is also the Director of the 'Center for Multivariate Analysis' there.

15.3 Special Honours and Awards

Professor C. R. Rao has the special distinction of being a Member of small group of scientists, who helped in the process of evolution of Statistics as an independent scientific discipline. It is now historically admitted that in the first half of the twentieth century, the foundations of Statistics were firmly rooted in powerful mathematical and probabilistic theories and techniques. B. Efron in his book titled 'The Statistical Century, Stochastic Musings: Perspectives from the Pioneers of the Late 20th Century' [Ed. John Panaretos, Lawrence Erlbaum Associates: New Jersey, (2003)] referred to that period as 'the golden age of statistical theory' and further wrote: 'Men of the intellectual calibre of Fisher, Neyman, Pearson, Hotelling, Wald, Cramer and Rao were needed to bring statistical theory to maturity.'

Prof. C. R. Rao being a world-famous scientist of top order received numerous awards and accolades. In 1967 he was elected a Fellow of the Royal Society, London. He was elected an Honorary Fellow of the Royal Statistical Society in 1969, the Calcutta Statistical Association in 1985, Finnish Statistical Society in 1990 and the Institute of Combinatorics and Applications in 1995. The same year, Prof. C. R. Rao was elected as a Member of the National Academy of Sciences, USA. He was elected an Honorary Member of the International Statistical Institute in 1983 and the International Biometric Society in 1986. He served as President of the Indian Econometric Society during 1971–1976, the International Biometric Society during 1973–1975, the Institute of Mathematical Statistics during 1976–1977, the International Statistical Institute during 1977–1979 and the Forum for Interdisciplinary Mathematics during 1982–1984.

Prof. C. R. Rao has been awarded 19 honorary doctorates from various Indian and foreign Universities. He has been the recipient of the Guy Medal in Silver of the Royal Statistical Society, London in 1965, the M. N. Saha Medal of the Indian

National Science Academy in 1969, the J. C. Bose Gold Medal and cash award in 1979, the Silver Plate bearing the monogram of the Andhra Pradesh Academy of Sciences in 1984, the S. S. Wilks Medal of the American Statistical Association in 1989 and the Mahalanobis Birth Centenary Gold Medal Award by the Indian Science Congress Association in 1996.

In 1968, the Government of India honoured Prof. C. R. Rao by conferring the 'Padma Bhushan' award on him.

In conclusion it may be pointed out that the two people who mentored Prof. C. R. Rao were always adulated by him as the 'guiding forces' of his life. One of them was Prof. P. C. Mahalanobis. Apart from being a legendary mathematician and statistician, he had the unique capability of recognizing the incredible talent of young C. R. Rao. It was he who guided C. R. Rao in the initial years of his career and brought out the best in him. The other mentor, in some sense, was Prof. R. A. Fisher. Next to Mahalanobis, Fisher was the greatest influence on Rao. As B. Efron wrote in his book published in 2003 [which has already been mentioned before]: 'Nobody was ever better at both the inside (mathematical foundations) and outside (methodology) of Statistics than Fisher, nor better at linking them together. His theoretical structures connected seamlessly to important applications and as a matter of fact, caused those applications to increase dramatically in dozens of fields.' On the occasion of Prof. Rao's 100th birthday, Efron wrote: 'Rao really was Fisher's student, in the sense of carrying on Fisherian statistical tradition'. Both Mahalanobis and Fisher were towering figures. Both of them were often uncompromising but extremely passionate about their work. They rightly and sincerely believed that practical applications should be the driving force behind statistical research. They had realized that applications often led to the development and discovery of new theories. Both had a penchant for the importance and correctness of collected data. Influenced by two such doyens of modern Statistics, C. R. Rao has himself become a formidable workforce in the field of modern-day Statistics.

15.4 Important Milestones in the Life of Professor Calyampudi Radhakrishna Rao

1920: Born on 10 September 1920 in a small town in the present-day state of Karnataka.

1940: Obtained the M. A. degree in Mathematics standing First with a First class from the Andhra University, Waltair.

1941: On 1st January 1941 joined the Indian Statistical Institute (ISI), Calcutta as a participant in an ongoing training programme in Statistics.

1941: In July 1941 joined as a student of the first batch of the M. A. course in Statistics at the University of Calcutta.

1943: Obtained the M. A. degree in Statistics from the University of Calcutta, standing First with a First class. He secured 87% of the total marks, a record still unbroken.

1943: Joined ISI as a 'Technical Apprentice' and also joined the Statistics Department of the Calcutta University as a part-time teacher.

1946: In August 1946 C. R. Rao joined the 'Anthropological Museum' of the Cambridge University as a full-time assistant to Prof. J. C. Trevor. Also enrolled at King's College, Cambridge, as a research student under the guidance of Prof. R. A. Fisher FRS.

1948: Obtained his Ph. D. degree in Statistics from the University of Cambridge,

1949: In July 1949, Dr. C. R. Rao joined the ISI, Calcutta, as a Professor and also as the Head of the famous 'Research and Training School' of ISI.

1949–1963: Served as Head of the 'Research and Training School' of ISI.

1963–1972: Served as Director of the 'Research and Training School' of ISI.

1953: 1953: His first Ph. D. student D. Basu obtained his degree in Statistics.

1965: Obtained the Sc. D. degree from the University of Cambridge for his overall contributions in Statistics. Awarded the Guy Medal in Silver from the Royal Statistical Society, London.

1967: Elected Fellow of the Royal Society (FRS), London. Awarded Doctorate degree (Honoris Causa) by Andhra University, Waltair.

1968: The Government of India conferred the 'Padma Bhushan' award on Prof. C. R. Rao.

1969: Elected Honorary Fellow of the Royal Statistical Society, London. Awarded the M. N. Saha Medal of the Indian National Science Academy.

Awarded Doctorate degree (Honoris Causa) by Leningrad University, USSR.

Served as President of the Indian Econometric Society.

Served as 'Chief Executive Secretary' of ISI. [The post was synonymous with Directorship]

1973: Awarded Doctorate degree (Honoris Causa) by Delhi University.

1973–1975: Served as President of the International Biometric Society.

1976–1984: Served as 'Jawaharlal Nehru Professor' at ISI.

1976: Awarded Doctorate degree (Honoris Causa) by University of Athens, Greece.

1976–1977: Served as President of the Institute of Mathematical Statistics.

1977: Awarded Doctorate degree (Honoris Causa) by Osmania University, Hyderabad.

1977–1979: Served as President of the International Statistical Institute.

1979: Awarded Doctorate degree (Honoris Causa) by Ohio State University, Columbus. Awarded the J. C. Bose Gold Medal and cash award.

1982: Awarded Doctorate degree (Honoris Causa) by Universidad Nacional de San Marcos, Lima, Peru.

1982–1984: Served as President of the Forum for Interdisciplinary Mathematics.

1983: Awarded Doctorate degree (Honoris Causa) by University of Philippines, Manila. Elected as Honorary Member of the 'International Statistical Institute'.

1984: Awarded the Silver Plate bearing the monogram of the Andhra Pradesh Academy of Sciences.

1985: Awarded Doctorate degree (Honoris Causa) by University of Tampere, Finland. Elected Honorary Fellow of the Calcutta Statistical Association.

1986: Elected as Honorary Member of the International Biometric Society.

1989: Awarded the S. K. Wilks Medal of the American Statistical Association. Awarded Doctorate degree (Honoris Causa) by Universite de Neuchatel. Awarded Doctorate degree (Honoris Causa) by the Indian Statistical Institute (ISI), Calcutta.

1990: Elected Honorary Fellow of the Finnish Statistical Society. Awarded Doctorate degree (Honoris Causa) by Colorado State University, Fort Collins, USA.

1991: Awarded Doctorate degree (Honoris Causa) by University of Hyderabad, Hyderabad. Awarded Doctorate degree (Honoris Causa) by Agricultural University of Poznan, Poland.

1994: Awarded Doctorate degree (Honoris Causa) by Slovak Academy of Sciences, Bratislava.

1995: Elected Member of the National Academy of Sciences, USA.

1995: Elected Honorary Member of the Institute of Combinatorics and Applications. Awarded Doctorate degree (Honoris Causa) by University of Barcelona, Spain. Awarded Doctorate degree (Honoris Causa) by University of Munich, Germany.

1996: Awarded Doctorate degree (Honoris Causa) by Sri Venkateswara University, Tirupati. Awarded Doctorate degree (Honoris Causa) by University of Guelph, Canada.

Chapter 16
Anadi Sankar Gupta (1932–2012)

16.1 Birth, Family, Education

Anadi Sankar Gupta was born on 1 November 1932 at Goila village of Barishal district in erstwhile East Bengal [present-day Bangladesh]. His father was Pramode Charan Gupta and mother Usharani Gupta. He had his school education at Domohani Kelejora High School in Asansol, West Bengal.

He had an illustrious academic career. In 1952 he obtained first class in Mathematics (Honours) from the prestigious Presidency College, Calcutta. In 1954, he passed the M. Sc. Examination with a first class from the famous Department of Applied Mathematics, University of Calcutta. In 1957, A. S. Gupta joined the Mathematics Department of the Indian Institute of Technology (IIT), Kharagpur in West Bengal, as an Assistant Lecturer. After joining the Institute along with his teaching assignments, he started pursuing research in Fluid Mechanics under the supervision of Prof. Gagan Behari Bandyopadhyay of the same Department. In 1959 he obtained his Ph. D. degree from IIT, Kharagpur, for his thesis titled 'Compressible flows with Heat Transfer and some Astrophysical applications', for original contributions in Fluid Mechanics. In 1967, he obtained D. Sc. degree in Mathematics for his thesis titled 'Stability and Heat Transfer in Fluid Flows'. This thesis was acclaimed for original contributions in 'Heat Transfer'.

16.2 Career, Research Contributions

In 1968 A. S. Gupta became a Professor in the Department of Mathematics, IIT, Kharagpur. He served there till 1993. Thereafter he served as INSA Honorary Scientist in the same Department till the end of his life. Prof. A. S. Gupta was considered as an outstanding teacher of Mathematics. At IIT, Kharagpur, he taught Mathematics and Mechanics to engineering students of undergraduate as well as Post-Graduate classes.

© The Author(s), under exclusive license to Springer Nature Singapore Pte Ltd. 2022
P. Mukherji, *Notable Modern Indian Mathematicians and Statisticians*,
https://doi.org/10.1007/978-981-19-6132-8_16

Apart from that he also taught many advanced topics in Mathematics, Mechanics and Fluid Mechanics to students of M. Sc. and Post-M. Sc. classes.

He made notable research contributions in Fluid Dynamics, Hydrodynamic and Hydro-magnetic stability, heat and mass transfer in fluid flows and Boundary Layer Theory. He made significant research work in the areas of both Newtonian as well as non-Newtonian Fluid Mechanics. Prof. A. S. Gupta initiated research in the areas mentioned above and built up a strong school of research in Fluid Mechanics at IIT, Kharagpur. B. S. Dandapath, L. Rai, S. Sengupta, P. S. Gupta, K. Rajagopal, R. N. Jana, N. Dutta and some others were his active collaborators. Prof. A. S. Gupta has made notable research in the areas of (i) Hydrodynamic Stability, (ii) Boundary Layer Theory, (iii) Heat Transfer and (iv) Magnetohydrodynamics (MHD). Some of his most significant contributions are being discussed here. His analysis of steady and transient-free convection in an electrically conducting fluid [Appl. Sci. Res. Holland, A9, (1960), 319] past a hot surface in the presence of a transverse magnetic field reveals that the field causes a reduction in the surface of heat flux. This result is in agreement with the experimental findings of A. F. Emery [Journ. Heat Transfer, U.S.A., Trans Amer. Soc. Mech. Engg., Ser C, (1963), 119]. Prof. A. S. Gupta and P. S. Gupta have also shown that homogeneous and heterogeneous chemical reaction result in a reduction of longitudinal (Taylor) diffusion coefficient of a solute dispersed in laminar channel flow [Proc. Royal. Soc., London, A 330, (1972), 59]. Prof. A. S. Gupta's corresponding analysis in the magnetohydrodynamic case [A. S. Gupta and N. Annapurna: Proc. Royal. Soc., London, A 367, (1979), 281] reveals that a uniform transverse magnetic field results in a reduction of Taylor diffusion coefficient of a solute dispersed in laminar flow of a conducting fluid in a channel.

During 1961–1962, Prof. A. S. Gupta went as a 'Senior Visitor' to the Department of Applied Mathematics and Theoretical Physics (DAMTP), University of Cambridge, England, under the 'Colombo Plan' scheme. There in collaboration with Prof. L. N. Howard [A. S. Gupta and L. N. Howard: Journ. Fluid Mech., 14, (1962), 463] Prof. Gupta derived general stability criteria for non-dissipative swirling flow of incompressible non-conducting as well as electrically conducting fluid permeated by magnetic fields. It is shown that for a swirling flow of a perfectly conducting fluid between two concentric circular cylinders in the presence of an axial current, a sufficient condition for stability with respect to axisymmetrical perturbations is that a suitable Richardson number modified to take account of the axial current is nowhere less than 1/4.

Prof. A. S. Gupta was awarded the prestigious S. S. Bhatnagar award in 1972 for his outstanding contributions in Fluid Dynamics and Magnetohydrodynamics. In the citation it was mentioned: 'He made significant contribution in the field of Fluid Dynamics and Magnetohydrodynamics notably on heat transfer in free convection flow in the presence of magnetic field. His work on the stability of a layer of rotating electrically conducting liquid in the presence of uniform magnetic field, oriented parallel to the axis of rotation is important, as it discusses the finite amplitude disturbance.'

During 1979–1981, Prof. A. S. Gupta went as a 'Visiting Professor' to the Department of Mechanical Engineering and Applied Mechanics, University of Michigan,

Ann Arbor, USA. There he worked in collaboration with Prof. C. S. Yih on the theory of laminar and turbulent buoyant plumes. They published the paper titled *'Plane buoyant plumes'* [Rev. Br. C. Mechanique, Rio di Janeiro, 111, (1981), 49].

Prof. A. S. Gupta's large number of investigations on the flow of Newtonian and visco-elastic fluids with particular reference to their flow over a stretching surface have important applications in the polymer processing and metal processing works. Stability of the flow of a viscous fluid over a stretching surface has been analysed by Prof. A. S. Gupta and S. N. Bhattacharya in the paper titled *'On the stability of viscous flow over a stretching sheet'* [Quarterly of Applied Mathematics, U. S. A., XLII, (1985), 359]. They showed that the flow was stable.

Prof. A. S. Gupta and S. N. Bhattacharya published another important paper titled *'Instability due to a discontinuity in magnetic diffusivity in the presence of magnetic shear'* [Journal Fluid Mechanics, 509, (2004), 125]. In this paper the linear stability of two viscous electrically conducting fluids separated by a plane interface and permeated by a sheared magnetic field parallel to the interface was studied. It was shown that if a magnetic field vanishes at the unperturbed interfaces, the configuration is always unstable for zero surface tension provided the magnetic diffusivities of the two fields are different.

Prof. A. S. Gupta has published a total of 157 research papers in reputed national and international journals. He authored the book titled 'Calculus of Variations with Applications' [Prentice Hall of India, New Delhi, 1997]. The important results from his research work have been cited in a large number of books and monographs. He supervised more than a dozen research scholars for their Ph. D. programmes.

16.3 Special Honours

For his original contributions in research in Applied Mathematics and for his excellence as a teacher, Prof. A. S. Gupta received several awards and was elected as 'Fellows' to learned bodies. It has already been mentioned that he received the prestigious 'Shanti Swarup Bhatnagar Award' in 1972. He was awarded the Federation of Indian Chambers of Commerce and Industry (FICCI) Award in 1978. In 1980 he was elected a 'Fellow' of the Indian National Science Academy (INSA). He was elected as the 'President' of the 'Annual Congress' of the 'Indian Society of Theoretical and Applied Mechanics' in 1985. In 1990 he was elected a 'Fellow' of the 'Indian National Science Academy (Allahabad)'. He received the 'P. L. Bhatnagar Memorial Lecture Award' in 1995. In 1999 he was elected the 'President' of the 'Annual Congress of Indian Mathematical Society'. Prof. A. S. Gupta served as an INSA Council Member during 2000–2002. He was the recipient of the 'Professor Vishnu Vasudeva Narlikar Memorial Lecture Award' of INSA in 2003. He became a Council Member of the 'Indian Mathematical Society' during the period 2007–2009.

He was a Member of the 'Editorial Board' of the Journal titled 'Stability and Applied Analysis of Continuous Media' (Italy). He was on the Editorial Board of 'Indian Journal of Pure and Applied Mathematics'.

In 2003, Prof. A. S. Gupta was made a 'Life Fellow' of IIT, Kharagpur, for his outstanding contributions and services to the Institute.

A modest, kind and simple man, Prof. A. S. Gupta will be long remembered for his remarkable research contributions, his excellent teaching abilities and the commendable school of research that he created with much care and commitment.

Prof. A. S. Gupta passed away in Kolkata on 14th June 2012 after a brief illness.

16.4   Important Events in the Life of Professor Anadi Sankar Gupta

1932: Born on 1st November 1932 in the district of Barishal in erstwhile East Bengal [present-day Bangladesh].

1952: Passed B. Sc. Examination with a first class in Mathematics (Honours) from the Presidency College, Calcutta.

1954: Passed M. Sc. Examination with a first class in Applied Mathematics from the University College of Science, Calcutta University.

1957: Joined the Department of Mathematics, IIT, Kharagpur, as an Assistant Lecturer in Mathematics.

1959: Obtained the Ph. D. degree in Applied Mathematics from the IIT, Kharagpur.

1961–1962: Visited the Department of Applied Mathematics and Theoretical Physics (DAMTP), University of Cambridge, England, as a 'Senior Visitor' under the 'Colombo Plan' programme.

1967: Obtained the D.Sc. degree in Applied Mathematics from the IIT, Kharagpur.

1968: Became a Professor in the Department of Mathematics, IIT, Kharagpur.

1972: Awarded the 'Shanti Swarup Bhatnagar' Award for original contributions in Fluid Mechanics and Magnetohydrodynamics.

1978: Recipient of the FICCI Award.

1979–1981: Went to Department of Mechanical Engineering and Applied Mechanics, University of Michigan, Ann Arbor, USA as a 'Visiting Professor'.

1980: Elected a 'Fellow' of the Indian National Science Academy (INSA), New Delhi.

1985: Elected 'President' of the 'Annual Congress' of 'Indian Society of Theoretical and Applied Mechanics'.

1990: Elected 'Fellow' of the 'National Academy of Sciences (India)' (Allahabad).

1995: Received the 'P. L. Bhatnagar Memorial Lecture Award' of the Indian Mathematical Society.

1999: Elected 'President' of the 'Annual Congress' of 'Indian Mathematical Society'.

2000–2002: Served as an INSA Council Member.

2003: Received the 'Professor Vishnu Vasudev Narlikar Memorial Lecture Award' of the Indian National Science Academy (INSA).

2003: Became a 'Life Fellow' of IIT, Kharagpur, West Bengal, India.

2007–2009: Served as a Council Member of the 'Indian Mathematical Society'.

2012: Died on 14th June 2012 in Kolkata after a brief illness.

Chapter 17
Jyoti Das (neè Chaudhuri) (1937–2015)

17.1 Birth, Education

Jyoti Chaudhuri, as she was known, prior to her marriage, was born on 21 August 1937 in a middle class Bengali family in Calcutta. She had her school education in Calcutta and passed Matriculation Examination in 1952 securing a first division. Then she joined the well-known Bethune College and completed her Intermediate Examination in Science (I.Sc.) in 1954 securing first division. She joined the Scottish Church College in Calcutta for pursuing her undergraduate studies. In 1956 Jyoti Chaudhuri obtained her B. Sc. degree standing First with a First class with Honours in Mathematics. She topped the list of all successful candidates in B. A. and B. Sc. Examinations of the University of Calcutta that year. She was awarded the 'Alfred Clarke Edward Scholarship', four gold medals, three silver medals and a book prize for her superlative scholastic performance. Then she joined the Department of Pure Mathematics, University of Calcutta, to continue her Post-Graduate studies. In 1958, she repeated her earlier performance, standing First with a First class in the M. Sc. Examination in Pure Mathematics and simultaneously topping the list of all successful candidates at the M. A. and M. Sc. Examinations of the University of Calcutta that year. She was once again awarded four gold medals and a book prize for her outstanding academic achievements.

17.2 Career, Research Contributions

After completing her Post-Graduate studies J. Das joined the Lady Brabourne College in Calcutta as a Lecturer in Mathematics. But soon she obtained the prestigious 'Commonwealth Scholarship' and proceeded to the University of Oxford, UK, to pursue research work in Pure Mathematics. In 1964, she was awarded the D. Phil degree by the University of Oxford for her original research on the 'Theory of Eigenfunction Expansions'.

© The Author(s), under exclusive license to Springer Nature Singapore Pte Ltd. 2022   137
P. Mukherji, *Notable Modern Indian Mathematicians and Statisticians*,
https://doi.org/10.1007/978-981-19-6132-8_17

Dr. J. Das had a long and illustrious teaching experience. After returning from Oxford University, in 1965, she took up a Faculty position in the Jadavpur University of Calcutta. After a brief stint there in 1966, she again left for the University of Dundee in Scotland. She taught and also continued her research work there. In 1968, she finally came back to India and joined the Indian Institute of Technology (IIT) in Madras. After 4 years, in 1972, she joined IIT, Kharagpur, in West Bengal. After serving the two premier IIT s in India for 7 long years, in 1975, she ultimately came back to Calcutta and joined her alma mater, the Department of Pure Mathematics of the University of Calcutta as a Reader. In 1979, she became the 'Sir Asutosh Birth Centenary Professor of Higher Mathematics' [previously known as Hardinge Professor] and thus became the first lady mathematician in India to hold a Chair Professorship in Mathematics. She worked in that capacity till the age of 65 and finally retired from the University of Calcutta in 2002.

Now the research contributions of Prof. J. Das are being discussed. Her research activities in the area of Pure Mathematics can be divided into the following four branches:

- Special Functions
- Eigenfunction Expansions
- Basics of Ordinary Differential Equations
- Some problems of Elementary Mathematics.

Prof. J. Das had started her research career, by working on 'Special Functions' under the guidance of her former teacher of Mathematics, Prof. Bholanath Mukherjee of Scottish Church College. Prof. Mukherjee was widely known all over India as the co-author of many text books in Mathematics for the undergraduate courses. Between 1962 and 1967 Prof. J. Das had published following six papers related to the above-mentioned topic. They are listed below:

- *'On Bateman-integral functions'* [Math. Zeitschrift, 78, (1962), 25–32]
- *'On a relation connecting the second solution of Tchebycheff's equations of the second kind and Bessel functions'* [Annali della Universita di Ferrara, X, (1962), 123–129]
- *'A note on definite integrals involving the derivatives of hypergeometric poly-nomials'* [Rendiconti del Seminario Mathematica della Universita di Padova, XXXII, (1962), 214–220]
- *'On the generalization of a formula of Rainville'* [Proc. Amer. Math. Soc., 17, (1966), 552–556]
- *'On the operational representation of some hypergeometric polynomials'* [Rendi-conti del Seminario Mathematica della Universita di Padova, XXXVIII, (1967), 27–32]
- *'Some special integrals'* [American Mathematical Monthly, 74, (1967), 545–548].

For forty years after these publications, Prof. J. Das concentrated on areas other than 'Special Functions'. But from 2007 onwards she again took up studies related to this area and published various research papers in 2007, 2008, 2011, 2012 and

2013. As these were all published in the twenty-first century, so they are beyond the purview of this monograph.

Before proceeding any further, it will be appropriate to write the difficulties she faced in starting her research career. It was not a smooth sailing by any means. With a commendable academic career, young J. Das enrolled herself as a research scholar at the Department of Pure Mathematics, University of Calcutta, towards the end of 1959, supported by a University Scholarship. Her registered supervisor was Dr. H. M. Sengupta, a renowned expert in the field of 'Mathematical Analysis'. Unfortunately, the scenario changed abruptly with the sudden demise of Dr. Sengupta. This also forced J. Das to give up the University Scholarship. In January 1961, she was appointed by the Public Service Commission, West Bengal, as a Lecturer in Mathematics at Lady Brabourne College, Calcutta. After a brief stint there, in October 1962, J. Das earned the Commonwealth Scholarship and proceeded to the University of Oxford in England, UK, to carry on research work in theory of 'Eigenfunction Expansions' under the guidance of an internationally famous mathematician Prof. E. C. Titchmarsh FRS, who was also the Sevilian Professor of Mathematics at the Oxford University. But ill-fate did not leave J. Das. Within three months of her arrival in Oxford University, Prof. Titchmarsh expired in January 1963. The young lady was then left with three options: (i) to return to India failing to carry on research under such an illustrious person like Prof. Titchmarsh; (ii) to enrol herself in some other University of UK; (iii) to continue with her research work in Oxford University under the guidance of Prof. J. B. Mcleod FRS, a former student of Prof. Titchmarsh. J. Das chose the third option as she was not willing to return to India without a doctoral degree from UK. But mentally she was rather upset because of past records of losing her supervisors.

At last she started working on a problem suggested by Prof. Mcleod. But J. Das failed to solve the problem and turned her attention to a problem on fourth-order ordinary differential equation and successfully completed the same. As the corresponding problem on second-order ordinary differential equation is not solvable, neither Prof. Mcleod nor Prof. W. N. Everitt was willing to accept J. Das's solution for the fourth-order ordinary differential equation, until she pointed out the advantage of the fourth-order problem over the second-order problem. Immediately the paper was sent to the Quarterly Journal of Mathematics and it was published in 1964. J. Das reported the same problem in the 'British Mathematical Colloquium' in 1963. J. Das completed her research works for the doctoral degree in a little more than a year and received the D. Phil degree from the University of Oxford in May 1964. She continued her research work in this field till 2004. During this period she published thirty-two research papers in the area of Eigenfunction Expansions and related topics. Some of these papers were written jointly with Prof. W. N. Everitt, a renowned mathematician and a Faculty Member of the Balliol College of Oxford University. Prof. Everitt was an expert in the field of Eigenfunction Expansion. In this context special mention needs to be made of the research paper titled '*On the square of a formally self-adjoint differential expression*' [Journ. London Math. Soc., (2), 1, (1969) 661–673]. This work was done and the paper written in collaboration with Prof. W. N. Everitt, and it opened up a new line of thought in this field. The paper determined the crucial clue necessary for the study of iterated second-order linear differential operators. Hundreds of papers have been written till date on the topic initiated by Professors J. Das and W. N. Everitt. A little history is associated with this problem. Dr. J. Das after coming back from Oxford in 1964 started working on the problem of the square of a formally self-adjoint second-order differential expression and communicated through letters on this work with Prof. W. N. Everitt. This correspondence continued for more than a year. Failing to arrive at a proper solution of the problem Prof. Everitt decided to sit together so as to resolve the problem fully. For this he arranged for Dr. J. Das a lecturer-ship at Queens College, Dundee, where he himself was working at that time. It was definitely an unprecedented move. Dr.J. Das was also so keen on solving the problem that she did not hesitate to resign from the post of a Reader in Mathematics, Jadavpur University, as she was not granted the required leave of absence without giving a bond for serving the University for three years after coming back. The story does not end here. Dr. J. Das joined Queen's College, Dundee, in October 1966, but the problem could not be solved till September 1967. By then, the Queen's College, Dundee, was upgraded to the status of the University of Dundee, and Dr. J. Das was offered permanent membership of the Faculty of the University. However, she refused the offer as she had the noble intention of serving her motherland.

In 1989, Prof. J. Das employed Co-ordinate Geometry to establish Weyl's limit classification of second-order ordinary differential equations. Normally, to analyse a nonlinear equation one considers a linear approximation of the given nonlinear equation and employs linear analysis to it to arrive at some result. The reverse process has never been encountered. In 1996 in her paper titled '*Nonlinear analysis as an aid to linear analysis*' [Journ. Pure Maths., 13, (1996), 1–12], Prof. Das showed that sometimes nonlinear analysis is useful in obtaining information about linear equation.

Towards the late nineties of the twentieth century, Prof. J. Das turned her attention to the basic theory of Ordinary Differential Equations. In 1998, in her paper titled '*A new method of solving linear homogeneous ordinary differential equations*' [Journ. Pure Maths., 17, (1998), 17–22], Prof. J. Das developed a new method of solving linear differential equations using functional analysis, which was a remarkable interdisciplinary work.

In the early part of the twenty-first century Prof. J. Das published a number of research papers (at least 6) where she developed various new methods and techniques for solving linear ordinary differential equations. But since this monograph has a specific timeframe, no detailed discussions of those papers are being made here. Again in two of her publications [2002, 2004], Prof. J. Das pointed out that there was much scope of doing research at the elementary level of Mathematics, like generalizations of Leibnitz's rule, Rolle's Theorem, etc. She also generalized 'Brahmagupta's triangles' and wrote a paper as a tribute to the renowned ancient Indian mathematician on his 1500th Birth Anniversary.

The complete list of Prof. J. Das's publication has been given along with the bibliographies of the other famous mathematical scientists. She published a total of 58 research papers in renowned international and national journals. She authored two textbooks. They are (i) 'Analytical Geometry' [Academic Publishers, Calcutta (2011) and (ii) 'Ordinary Differential Equations' [Academic Publishers, Calcutta (2015)]. The last book was published after the death of Prof. J. Das.

Eight researchers obtained their Ph. D. degrees under the guidance of Prof. J. Das. Prof. V. Krishna Kumar and Prof. Jayasree Sett are two of her well-known students; both of them became Professors in different Universities/national institutes. While the first six of her Ph. D. students worked on the 'Theory of Eigenfunction Expansions', the seventh one worked on a fundamental question of 'Special Function'. The topic of the dissertation of the last student was remarkably on an innovative idea associating geometrical interpretations with subsets of the solution set of given first-order partial differential equations.

Apart from doing serious research in Mathematics, Prof. J. Das also devoted her time for popularizing Mathematics among school and college students. She wrote several articles and rhymes in Bengali on Mathematics.

She was the greatest lady mathematician not only in Bengal but perhaps in the whole of Eastern India. She served in many important national and state-level academic committees.

Prof. J. Das was a brilliant academician, a remarkable researcher, an outstanding teacher and a very sincere and caring research supervisor. Just as her students used to adore her, she too loved them with maternal grace. In personal life, she was a dutiful wife and a loving mother and a very modest person. She expired in Calcutta after a brief illness on 21st May in 2015.

17.3   Special Honours

Prof. J. Das was a Founder Fellow of the 'West Bengal Academy of Science and Technology' (WAST). The University of Calcutta bestowed on her the 'Outstanding Faculty' award for her long years of remarkable service to the University and also for her path-breaking original research work in her fields of specialization.

17.4   Important Milestones in the Life of Professor Jyoti Das

1937: Born in Calcutta on 21st August 1937.

1956: Obtained the B.Sc. degree in Mathematics (Honours) standing First with a First class from the Scottish Church College, Calcutta. She topped the list of all successful candidates in B. A. and B. Sc. Examinations of the University of Calcutta that year.

1958: Obtained the M. Sc. degree in Pure Mathematics standing First with a First class from the University of Calcutta. She topped the list of all successful candidates in M. A. and M. Sc. Examinations of the University of Calcutta that year.

1961: Joined the Lady Brabourne College in Calcutta as a Lecturer in Mathematics.

1962: Joined the University of Oxford, England, UK as a 'Commonwealth Scholar' to pursue higher studies.

1964: Obtained the D. Phil degree from the University of Oxford for her original research work on 'Theory of Eigenfunction Expansions' under the joint supervision of Professors J. B. Mcleod FRS and W. N. Everitt.

1964: Joined the Department of Mathematics, Jadavpur University, Calcutta, as a 'Reader'.

1966: Joined the Queens' College, Dundee, Scotland, as a Lecturer in Mathematics.

1975: Joined the Department of Pure Mathematics, University of Calcutta, as a 'Reader'.

1979: Became the 'Sir Asutosh Birth Centenary Professor of Higher Mathematics' (previously known as Hardinge Professor) and also the Head of the Department of Pure Mathematics, University of Calcutta. She became the first lady mathematician in India to adorn a Chair Professorship in Mathematics.

2002: Retired from the University of Calcutta.

2015: Passed away in Calcutta after a brief illness on 21st May 2015.

Chapter 18
Some Outstanding Minds
from the Indian Statistical Institute

In this chapter, some brief discussions will be made on 7 renowned researchers who were trained in ISI and/or University of Calcutta and while serving ISI made outstanding contributions in the areas of Mathematics, Statistics and Probability. They left ISI and migrated to USA. But still, they were very much a part of ISI in the second half of the twentieth century. Their remarkable contributions in their respective fields of research added to the glory of ISI. The names of these 7 great personalities have been arranged chronologically.

18.1 Samarendra Nath Roy [S. N. Roy] (1906–1964)

S. N. Roy was born on 11 December 1906 in Calcutta. His father Dr. Kali Nath Roy was a famous journalist and the editor of the daily newspaper '*The Tribune*', which was published from Lahore, in pre-independent undivided Panjab. His mother Suniti Bala Roy was a housewife.

Right from the early school days, he was a very bright student of Mathematics and had a uniformly brilliant academic career. In 1928, he stood First in Calcutta University with a First class Honours in Mathematics in the B.Sc. Examination from the prestigious Presidency College of Calcutta. In 1931 he again stood First with a First class in the M.Sc. Examination from the Department of Applied Mathematics, University of Calcutta. While studying in the Post-Graduate class, he had chosen 'Theory of Relativity' as his special paper. After completing the master's degree he initially started doing research on cosmological problems under the supervision of Prof. N. R. Sen. At that time S. N. Roy used to go to the newly established 'Indian Statistical Institute' at Presidency College to use the computational facilities available there. During such visits, S. N. Roy met Prof. P. C. Mahalanobis. Impressed by his mathematical acumen, Prof. Mahalanobis persuaded Roy to join the small group of researchers at ISI. In 1934, S. N. Roy joined as a part-time researcher, and a year later, he was recruited as a full-time statistician. S. N. Roy has made outstanding

P. Mukherji, *Notable Modern Indian Mathematicians and Statisticians*,
https://doi.org/10.1007/978-981-19-6132-8_18

contributions in the area of 'Multivariate Analysis' in Statistics. Special mention needs to be made of the paper titled '*Distribution of the studentized D^2 statistic*' [(With R. C. Bose) Sankhyā, 4 (1938), 19–38]. In this research paper, R. C. Bose and S. N. Roy jointly worked out the non-null distribution of the studentized D^2-statistic, which was being used by Prof. P. C. Mahalanobis as an important tool in his anthropometry-related studies. This new distribution facilitated the computation of power function of Hotelling's T^2-statistic. The very next year he published a paper titled '*p-statistics or some generalizations in analysis of variance appropriate to multivariate problems*' [Sankhyā, 4 (1939), 381–396]. This is another notable contribution, in which S. N. Roy fully worked out the sampling distributions of *p*-statistics.

While still being actively involved in statistical research in ISI, in 1938, S. N. Roy was appointed as a Lecturer in his alma mater, the Department of Applied Mathematics of Calcutta University. In 1941, S. N. Roy was transferred to the newly created Department of Statistics of the same University. There he helped Prof. P. C. Mahalanobis and R. C. Bose to build up the new Department of Statistics. From 1946 to 1949, he served as the Assistant Director of ISI. Along with this in 1947–1948, S. N. Roy acted as the temporary Head of the Department of Statistics. In spite of such heavy administrative responsibilities, S. N. Roy continued to do notable research in discrete and continuous Multivariate Analysis. He published another important paper titled '*The sampling distribution of p-statistics and certain allied statistics on the non-null hypothesis*' [Sankhyā, 4 (1942), 15–34]. He derived the non-null distribution of the latent roots. A related paper titled '*The individual sampling distribution of the maximum and minimum and any intermediate one of the p-statistics on the null hypothesis*' [Sankhyā, 7 (1945), 133–158] is also important. Around that time he wrote a series of papers on statistical inference in multivariate problems.

He did his doctorate under the joint supervision of Professors P. C. Mahalanobis and N. R. Sen. During 1946–1949, Dr. S. N. Roy served as the Assistant Director of ISI, Calcutta. In spite of heavy class loads and administrative duties, in collaboration with Prof. P. C. Mahalanobis and R. C. Bose, he made notable contributions in different areas of multivariate analysis, *p*-statistics, etc.

In the spring of 1949, Dr. S. N. Roy went to USA as a Visiting Professor of Statistics at Columbia University. After he returned to Calcutta, he was made the Head of the Department of Statistics at the University of Calcutta. However, in March, 1950 he left India permanently and joined the Department of Statistics at the University of North Carolina at Chapel Hill as Professor of Statistics.

During his stay in USA, he worked on many areas of Statistics such as analysis of contingency tables, distribution of eigen values and eigen vectors, growth curve analysis, introduction of a heuristic principle of construction of tests of hypotheses, based on the union and intersection principle.

He wrote many important research papers and two useful monographs. The first monograph is titled 'Some Aspects of Multivariate Analysis' [John Wiley and Sons, New York (1957)] and the second one is titled 'Analysis and Design of Certain Quantitative Multi-response Experiments' [S. N. Roy, R. Gnanadesikan and J. N.

Srivastava, Oxford, New York, Pergamon Press (1971)]. However due to his sudden demise in July 1964 the second book was published posthumously.

Prof. S. N. Roy received many academic honours for his original contributions in Statistics. In 1946, he was elected Fellow of the National Institute of Sciences in India [now known as 'Indian National Science Academy']. He was President of the Statistics Section of the Indian Science Congress held at Patna in 1948. In 1951, Prof. S. N. Roy was elected a Fellow of the International Statistical Institute.

Prof. S. N. Roy passed away in July 1964 while taking a holiday in Canada.

18.2 Debabrata Basu [D. Basu] (1924–2001)

D. Basu born in Dacca in erstwhile East Bengal (present-day Bangladesh) on 5 July 1924. His father N. M. Basu was a well-known mathematician of his time and did some notable work on Number Theory. D. Basu had his early education in East Bengal and obtained his master's degree from the Dacca University in 1947. In 1948, after the partition of India, he came to Calcutta permanently. He worked for a brief period in a private Insurance Company. In 1950, he joined ISI, Calcutta, as a research scholar under the supervision of Prof. C. R. Rao. This was a turning point in the life of D. Basu. A famous Hungarian mathematician Abraham Wald (1902–1950) was visiting and delivering lectures on Bayesian Statistics at ISI, Calcutta, at that time. Incidentally Wald was quite well known for his research contributions on Decision Theory, Geometry and Econometrics. D. Basu attended his lectures and was highly impressed by Wald.

In 1953, D. Basu obtained his Ph.D. degree in Statistics from the University of Calcutta. He was the first student of Prof. C. R. Rao to obtain the Ph.D. degree under his guidance. Shortly after that D. Basu obtained the Fulbright Scholarship and went to the University of California, Berkeley, USA. There he came in close contact with renowned statisticians such as Jerzy Neyman (1894–1981) and E. L. Lehman (1917–2009).

D. Basu is noted for inventing simple examples that are responsible for displaying some difficulties of likelihood-based Statistics and frequentist Statistics. His paradoxes are considered to be especially important in the development of Sample Survey. Interestingly, according to D. Basu, he had long discussions with Prof. R. A. Fisher for finding out a middle course between Bayesian Statistics and the Statistics as practised at the Berkeley School. This ultimately led to his noted theorem which has established the independence of complete sufficient Statistics and ancillary Statistics.

Dr. D. Basu taught at ISI, Calcutta, for sometime but then he moved over to other places. In 1975, he permanently left for USA, where he taught at the Florida State University for 15 years. In 1979, he was elected a Fellow of the 'American Statistical Association'. He was also a Fellow of the 'Institute of Mathematical Statistics' and an elected Member of the 'International Statistical Institute'. Six students received their doctorate degrees under the supervision of Prof. D. Basu.

Prof. D. Basu passed away in Calcutta on 24 March 2001.

18.3   R. Ranga Rao (1935–2021)

R. Ranga Rao was born in 1935 in the city of Madras (present-day Chennai). After obtaining his B.Sc. in Mathematics (Honours) from the University of Madras, Ranga Rao came to Calcutta. In 1957, he joined the ISI, Calcutta, as a research scholar under the guidance of Prof. C. R. Rao. He was one of the 'group of 4', referred to as 'famous four' in ISI during 1956–1963. He obtained his Ph.D. degree from the University of Calcutta. Shortly after that he permanently migrated to USA and joined the University of Illinois at Urbana-Champaign, USA. He served there for 40 long years and retired in 2001. Prof. Ranga Rao is renowned for his contributions in Lie Groups and Lie Algebra. His name is associated with the famous Kostant–Parthasarathy–Ranga Rao–Varadarajan determinants published in 1967. He returned to his home town Chennai in 2015. In 2021 he passed away in the same city.

18.4   Kalyanapuram Rangachari Parthasarathy [K. R. Parthasarathy] (Born 1936)

K. R. Parthasarathy was born on 25 February 1936 in the city of Madras (present-day Chennai). He completed his B.A. in Mathematics (Honours) from the Ramakrishna Mission Vivekananda College under the University of Madras. He joined the ISI, Calcutta, sometime in the late fifties of the last century as a research scholar under Prof. C. R. Rao. He completed his Ph.D. in 1962 and was the first recipient of the doctorate degree of ISI. He received the Shanti Swarup Bhatnagar Prize in Mathematical Sciences in 1977. His field of specialization is 'Quantum Probability'. He is also the recipient of the 'Third World Academy of Science' (TWAS) Prize for his outstanding contributions in Mathematical Sciences. After completing his Ph.D. from ISI, Dr. K. R. Parthasarathy worked during 1962–1963 as Lecturer at 'Steklov Mathematical Institute' of the USSR Academy of Sciences. There he was fortunate to work in collaboration with the world-renowned Soviet mathematician Prof. Andrey Kolmogorov (1903–1987). As is well known, Prof. Kolmogorov is internationally acclaimed for his remarkable contributions in the fields of Mathematics of Probability, Topology, Intuitionistic Logic, Computational Complexity and algorithm Information Theory. During 1964–1968, Dr. K. R. Parthasarathy served as a Professor of Statistics in the University of Sheffield. During 1968–1970, he worked at the University of Manchester and later at the University of Nottingham. At the University of Nottingham, Dr. K. R. Parthasarathy collaborated with noted British mathematician Prof. R. L. Hudson (1940–2021) and did pioneering work in 'Quantum Stochastic Calculus'. After spending a few years at Bombay University and IIT, Delhi, in 1976 Dr. Parthasarathy joined as a Professor in the Department of Statistics of the new Indian Statistical Institute, Delhi Centre. He served there till his retirement in 1996.

K. R. Parthasarathy was one of the 'group of 4', referred to as 'famous four' in ISI during (1956–1963). His name is associated with the famous Kostant–Parthasarathy–Ranga Rao–Varadarajan determinants published in 1967.

Prof. Parthasarathy has authored two books titled:

- Probabilistic Measures on Metric Spaces [Vol. 252, American Mathematical Society (1967)]
- An Introduction to Quantum Stochastic Calculus [Vol. 85, Springer (1992)].

18.5 Jayanta Kumar Ghosh [J. K. Ghosh] (1937–2017)

J. K. Ghosh was born in Calcutta on 23 May 1937. He obtained his undergraduate degree in Statistics from the Presidency College, Calcutta. He got his M.Sc. in Statistics from the University of Calcutta. He started doing research under the guidance of Prof. H. K. Nandi of the Statistics Department of Calcutta University. His area of specialization was Bayesian Mathematical Statistics and applications. Early in his career he started studying sequential analysis. He worked on invariance, sufficiency and applications to sequential analysis. He obtained his Ph.D. degree in Statistics from the University of Calcutta. In his Ph.D. thesis, he studied average sample number and the efficiency of sequential t-test. He was influenced by Abraham Wald's invention of the Sequential Probability Ratio Test (SPRT). He continued his research related to these topics and was able to prove under very general conditions that in reducing a problem through sufficiency and invariance, the order of applications of these criteria is immaterial. He is supposed to have made significant contributions to theoretical Statistics in various directions. He extended the results of Fisher and Rao on the second-order efficiency in maximum likelihood estimators. By making use of refined analytical tools, Prof. J. K. Ghosh obtained useful results in the asymptotic expansion of the distribution of sample Statistics. As regards applications of probabilistic methods, he made notable contribution to the understanding of sediment transport in fluid flows through stochastic model. He received the prestigious Shanti Swarup Bhatnagar Award in Mathematical Sciences in 1981. He was elected a Fellow of the Indian National Science Academy in 1982. He served the Indian Statistical Institute, Calcutta, for many years and became the Director of the Institute in the mid-eighties of the twentieth century. He has published more than 50 research papers and supervised 25 doctoral students. However, soon after his retirement from ISI, he migrated to USA and joined the Purdue University there. In 2014, the Government of India honoured him with the 'Padma Shree' Award. At the age of 80, he passed away in 2017 in the USA.

18.6   Veeravalli Seshadri Varadarajan [V. S. Varadarajan] (1937–2019)

[During proof reading incorporate corrected years]

V. S. Varadarajan was born on 18 May 1937 in Bangalore (present-day Bengaluru). As his father Seshadri Varadarajan worked as an Inspector of Schools, he was transferred to various places. V. S. Varadarajan had his entire early school education at Tiruchirappalli and Salem. When he finished high school, his father was transferred to Madras (present-day Chennai). So he studied there, and in 1954 he completed his I.Sc. Examination from the famous Loyola College of Madras. In 1956, he obtained his B.Sc. degree with Statistics (Honours) from the Presidency College under the University of Madras. But the course contained a lot of Mathematics as well. During his final year at Presidency College, Varadarajan had the privilege of listening to a lecture delivered by legendary Prof. C. R. Rao. That inspired him and so he came to Calcutta and in 1956 joined the ISI, Calcutta, as a research scholar under renowned statistician Prof. C. R. Rao. In a courageous move, V. S. Varadarajan started working on Probability Theory rather than in Statistics. He obtained his Ph.D. under the guidance of Prof. C. R. Rao in 1960 from the University of Calcutta. Shortly after that he went to the famous Princeton University as Post-doctoral fellow. There he met the well-known American mathematician Victor Bargmann (1908–1989), who inspired him to study the works of the famous Indian-American mathematician Harish-Chandra (1923–1983). Towards the end of 1960, Dr. V. S. Varadarajan went to the University of Washington, Seattle, and spent an academic year there. Following that, he spent a year at the famous 'Courant Institute' at the New York University. During his stay at USA he became interested in 'representation theory'.

Thereafter, in the later part of 1962, Dr. Varadarajan returned to ISI, Calcutta. With fresh ideas in his mind, he encouraged his friends Ranga Rao and Parthasarathy to study the works of Soviet mathematicians I. Gelfand (1913–2009) and M. A. Naimark (1909–1978) and of course Harish-Chandra. That opened up new avenues of research for the young scholars.

After spending about three years in his alma mater, the ISI, Calcutta, in 1965, Dr. V. S. Varadarajan permanently migrated to USA. There he served at the University of California, Los Angeles, till his retirement.

It may be noted that Prof. Varadarajan has made commendable research contributions in the areas of Probability Theory, Lie Groups and their representations, Quantum Mechanics, Differential Equations and Supersymmetry. Some of his best known work is in the 'representation theory' and related topics. Along with the leading American mathematician Bertram Kostant (1928–2017), who was considered a leading figure in 'representation theory', Prof. Varadarajan introduced the famous 'Kostant–Parthasarathy–Ranga Rao–Varadarajan' determinant in 1967. His other two notable contributions are 'Trombi–Varadarajan Theorem' established in 1972 and 'Enright–Varadarajan modules' discovered in 1975. He was awarded the 'Onsager Medal' for his original contributions in Mathematical Sciences.

Along with R. Ranga Rao, K. R. Parthasarathy, S. R. S. Varadhan, V. S. Varadarajan was one of the 'group of 4', referred to as 'famous four' in ISI during (1956–1963). During the twentieth century he published several important books in advanced areas of Mathematical and Physical Sciences. They are:

- Geometry of Quantum Theory [Springer Verlag (1968)]
- Lie Groups, Lies Algebras and Their Representations [Springer Verlag (1974)]
- Harmonic Analysis on Real Reductive Groups [Springer Verlag (1977)]
- An Introduction to Harmonic Analysis on Semisimple Lie Groups [Cambridge University Press (1989)].

He was a diabetic for a long time, and finally on 25 April 2019, the outstanding mathematician Prof. V. S. Varadarajan passed away at California, USA.

18.7 Sathamangalam Ranga Iyengar Srinivasa Varadhan FRS [S. R. S. Varadhan] (Born 2 January 1940)

S. R. S. Varadhan was born on 2 January 1940 in Madras. He obtained his under-graduate degree in Mathematics from the Presidency College under the University of Madras. In 1959, S. R. S. Varadhan joined ISI as a research scholar under Prof. C. R. Rao. Initially, the young enthusiast was desirous of working in Statistical Quality Control. But the story goes, that after going through the well-known book on Measure Theory by Halmos, Varadhan changed his mind and joined the famous group of researchers on Probability Theory, referred to as 'famous four' in ISI during (1956–1963). One important Member of the group V. S. Varadarajan was however away to USA on a Post-doctoral fellowship at that time. He received his Ph.D. degree from ISI in 1963 for his thesis titled 'Convolution Properties of Distributions on Topological Groups'. Prof. C. R. Rao had arranged for the famous Soviet mathematician Prof. A. Kolmogorov to be present at the defence of Varadhan's Ph.D. thesis.

Soon after that Dr. S. R. S. Varadhan migrated to USA. From 1963 to 1966 he worked as a Post-doctoral Fellow at the 'Courant Institute of Mathematical Sciences' at New York University. After that he joined the Institute as a Faculty Member and has been there since then. Prof. Varadhan is internationally famous for his original contributions in the Probability Theory and especially for creating a unified theory of large deviations.

Though he left ISI, Calcutta, and India permanently in 1963, but still he is undoubtedly one of the most outstanding mathematicians of his time who was trained up at the institute. He has won numerous awards, including the coveted 'Abel Prize' in 2008 for his fundamental contributions on large deviations. The Government of India has honoured him with 'Padma Bhushan' Award in 2007.

Epilogue

This chapter has two sections in it. In the first section 'Concluding Remarks', a brief discussion has been made mainly on the general level of research activities in Mathematics and Statistics (including Probability and other related areas) in the two giant institutions, namely the 'University of Calcutta' and the 'Indian Statistical Institute', excluding the contributions of the pioneers already discussed in details. The academic impacts of these institutions in the national and international context have been mentioned. Their current status has also been discussed.

The second section 'Compiled Bibliographies of 16 Listed Mathematical Scientists' is devoted exclusively to the bibliographies of the 16 great pioneers in their respective fields in Mathematics/Statistics. But in the case of two such personalities, namely Prof. R. C. Bose and Prof. C. R. Rao, the publications made by them during their stay in Bengal (formally) only have been included. As both of them finally left India and immigrated permanently to USA, the cut-off is essential, keeping in mind the fact that the monograph relates to Bengal only. Secondly publications made beyond the year 2000 A.D. have not been considered because of the time-constraint of the monograph (nineteenth and twentieth centuries).

Concluding Remarks

The Current Status of the Institutions

First a brief discussion about the research activities in the two Departments of Mathematics in Calcutta University is necessary. If an unbiased and neutral observation is made then one can safely say that both in the Departments of Pure Mathematics as well as Applied Mathematics of the 'University of Calcutta', the research activities after the seventies of the twentieth century were at the best sporadic.

Right from its inception, the Department of Pure Mathematics had a strong culture of research on Geometry of various types. This has been discussed in details while dealing with pioneers Syamadas Mukhopadhyay, R. N. Sen and R. C. Bose. They made memorable contributions and were acclaimed internationally. The departure of R. C. Bose from the department was a great loss. M. C. Chaki did make some valuable additions and also trained up a number of students in these areas. But unfortunately, the quality of research in Geometry fell to mediocrity and lost its earlier level of excellence.

During the early part of the twentieth century, none in this department did any serious work either in the field of Algebra or Mathematical Analysis. Strangely enough, though the first Hardinge Professor of Higher Mathematics Prof. W. H. Young FRS was himself an internationally famous researcher in 'Mathematical Analysis', he was not able to train up anyone in this discipline. However, Prof. F. W. Levi, who had to leave Germany because of the anti-Semitic policies of the then Nazi rulers accepted an offer and joined the Department of Pure Mathematics as the Hardinge Professor and Head of the Department in 1935. He played a very important role by introducing modern mathematical disciplines such as 'Abstract Algebra', 'Set Theory', 'Combinatorics', 'Topology' in the syllabus of Calcutta University. It was Prof. Levi, who played a very important role in the academic life of Prof. R. C. Bose. In a sense, he was the man who took the initiative to upgrade the standard of teaching in the Department of Pure Mathematics on par with international Universities.

H. M. Sengupta who came to the Department of Pure Mathematics of Calcutta University from Dacca during the late forties of the twentieth century tried to build a research school on 'Mathematical Analysis' and the famous number theorist from South India. Prof. S. S. Pillai (1901–1950) who joined the same department at the invitation of Prof. Levi tried to train up students on 'Number Theory'. But sudden and untimely demise of both of them were irreparable losses to the department.

Prof. (Mrs.) Jyoti Das was herself both a brilliant academician and a researcher. Her contributions have been discussed in details in Chap. 17. But unfortunately, she could not build up proper school in Differential Equations, Eigenfunction Expansions and allied fields.

However, during the thirties, forties and fifties of the twentieth century the Department of Pure Mathematics of Calcutta University was famous for teaching and research throughout India. Many bright students from other states came and studied in this department for the Post-Graduate course. They also carried out research and obtained their doctorate degrees from the University of Calcutta. The most famous among them was R. C. Bose. His life and work have been elaborately discussed in Chap. 11. Three noted mathematicians of the state of Karnataka pursued their Post-Graduate studies in Calcutta University. Prof. B. S. Madhava Rao (1900–1987) obtained his master's degree from the Department of Applied Mathematics, but he also specialized in various branches of Pure Mathematics too. He also obtained the D.Sc. degree from the University of Calcutta. Later he became a famous Faculty Member of Central College, Bangalore. He was a well-known mathematician throughout India. Prof. K. Venkatachaliengar (1908–2003) pursued his Post-Graduate studies and stood First with a First class from the Department of Pure

Mathematics. He also obtained the D.Sc. degree from the Calcutta University for his contributions in Pure Mathematics. He too was a well-known mathematician in the state of Karnataka as well as India. Prof. S. V. Hegde (1922–1976) also did his M.Sc. from the Department of Pure Mathematics, University of Calcutta. He died rather early, and the state of Karnataka lost a flourishing talented mathematician.

The Department of Pure Mathematics suffered a gradual decay for various reasons. The appointments of Faculty Members were often biased, and the University administration showed no interest in developing essential infrastructures such as a good library with modern facilities, lack of adequate state-of-the-art computational availability, no programmes such as exchange of Faculty with top mathematical institutions of India such as TIFR, HRI, IMSc were undertaken, nor were the syllabi updated or new disciplines introduced to keep up the standard on par with other centres of excellence. The ultimate outcome was the gradual downfall of an erstwhile great centre of teaching and research in the area of Pure Mathematics.

The downfall of the legendary Department of Applied Mathematics was equally pathetic. After the retirement of Prof. N. R. Sen, the said department virtually started suffering from an identity crisis. The appointments of Faculty Members with questionable academic accomplishments and lack of initiative to upgrade the contents of the curriculum were two major lapses. After Prof. S. K. Chakrabarty took charge of the department in 1963, he did try to implement some corrective measures. He was instrumental in appointing some talented young men as Members of the Faculty; he also updated the syllabus keeping in mind the requirements of the time. He was fully aware about the need to install computers in the department to facilitate computer-related studies and research as well as computational facilities. The IBM had agreed to install a computer in the campus of the University College of Science [Rajabazar Campus], mainly because of Prof. Chakrabarty's relentless efforts. But political interference from various 'unions' in the University, supported by the then ruling party hierarchy in West Bengal, was responsible for stopping the installation. This was a major setback for the department. Undaunted, Prof. S. K. Chakrabarty worked untiringly and somehow managed to include the theoretical study related to computer hardware and software in the Post-Graduate syllabus of Applied Mathematics. But lack of computers was a great hindrance to the thorough learning of the subject. So the department lost the race to become a top-grade centre of computer-oriented studies in the country. Some people did carry on some research in Solid, Fluid Mechanics and Mathematical Physics. But the outcome was nothing remarkable.

However, it should be noted that this department too attracted many bright students from other states during its hay day. In this context Prof. B. S. Madhava Rao has already been mentioned. Prof. Phoolan Prasad (born 1944) was born in the state of Bihar. But he pursued his graduate and Post-Graduate studies in the famous Presidency College and the Department of Applied Mathematics, University of Calcutta, respectively. After that he shifted to the Indian Institute of Science (IISc) in Bangalore and completed his doctoral work there. Later he became a famous Faculty Member of the Department of Mathematics of IISc, Bangalore. He is a leading applied mathematician of India, his area of specialization being Fluid Mechanics and related areas.

Another problem, which this department faced right from the thirties of the twentieth century, was exodus of brilliant students to greener pastures. Some went to ISI, Calcutta. Many went abroad. Especially, after India gained her independence the number of students going to USA in large numbers dealt a grievous blow to department. Thus a department once considered as a centre of excellence in Applied Mathematics became a very mediocre centre of teaching and research.

Now an assessment of the research activities in mathematical sciences at the Indian Statistical Institute (ISI) is necessary too. As has already been discussed in the first two chapters, under the dynamics and innovative leadership of Prof P. C. Mahalanobis, a strong culture of research in Statistics prevailed at the ISI right from the time of its inception in the early thirties of the last century. From 1935 onwards research on mathematical disciplines such as Combinatorics, Linear and Abstract Algebra started mainly because of Dr. R. C. Bose. Then in the forties, R. C. Bose along with C. R. Rao collaborated with the famous number theorist and combinatorial mathematician Prof. S. S. Chowla then based in Lahore. Mathematical research on a wide spectrum of areas was a hallmark of ISI right from then. Under the leadership of two of the century's greatest statisticians Professor P. C. Mahalanobis and C. R. Rao, research on both theoretical and applied Statistics continued with rapid strides. Even though Prof. Mahalanobis was not very keen on 'Mathematical Statistics', the culture of studies in that area spread to ISI from the early fifties. The visit of Abraham Wald created a lot of excitement at the institute. D. Basu (about whom some discussions have been made in Chap. 18) was the first man to start studies and research on Mathematical Statistics. The visit of famous statisticians from abroad and visit of many junior faculties of the institute such as S. K. Mitra, R. G. Laha, J. Roy to USA for their Ph.D. research or brief visits strengthened the development of research in Mathematical Statistics at ISI. The research output was of high standard. By mid-fifties the institute had an excellent Faculty who were carrying on research in many branches of Statistics including Mathematical Statistics. But research activities in Probability Theory and stochastic processes were still lagging behind.

The 1956 was a year which will always have a special place in the calendar of ISI. A young student V. S. Varadarajan after completing his graduation from Madras University and joined ISI that year as a research scholar under Prof. C. R. Rao. Instead of working in Statistics, he decided to work in the theory of Probability. Very courageously, he started learning modern Probability Theory and the necessary Mathematics all on his own. Soon two other students from Madras, R. Ranga Rao and K. R. Parthsarathy completed their graduation from Madras University and joined ISI, again under Prof. C. R. Rao. They became very close to Varadarajan and chalked out an ambitious programme of study for themselves. They learnt what they could from the existing Faculty Members of ISI. But apart from that, they worked very hard with great care and thoroughness and were able to reach out and understand some of the sophisticated frontier areas of theory of probability. In 1959, Varadarajan was able to submit his thesis on convergence of probability measures on metric spaces. That same year another brilliant youngster followed the same route from Madras. After completing his graduation from Madras University, he came and joined ISI as a research scholar of Prof. C. R. Rao. He too became very close to his predecessors from Madras and soon they came to be known as the 'group of four' at the institute. This small and extremely dedicated group of researchers continued their vigorous research work, and their interests went through a variety of topics such as topological and analytical aspects of probability theory, probability theory of groups, ergodic theory, mathematics of quantum theory and representation theory of groups. These four outstanding scholars have been discussed in Chap. 18, titled 'Some Outstanding Minds from the Indian Statistical Institute'. Their contributions in later life made the name of ISI famous throughout the world.

Thus ISI which was the fulfilment of the dream of a genius scientist Prof. P. C. Mahalanobis scaled glorious heights during his own lifetime. The brilliance and committed dedication of Prof. C. R. Rao carried forward the flame, and the institute still continues to be a centre of excellence in teaching and research not only in Statistics and Mathematics but in various other branches of sciences as well. It should be specially mentioned that not only in India, but all over the world, students taught and trained at ISI, Calcutta (Kolkata), have done very well professionally and are still considered to be among the best. Many mathematicians and statisticians who were students of ISI, Calcutta (Kolkata), have adorned prestigious positions both in international academic institutions as well as international administrative agencies such as the UNESCO, FAO and UNICEF.

Compiled Bibliographies of 16 Listed Mathematical Scientists

List of Publications of Sir Asutosh Mookerjee

1. Proof of Euclid I, 25 [Messenger of Mathematics, 10, (1880–1881), 122–123].
2. Extension of a theorem of Salmon's [Messenger of Mathematics, 13, (1883–1884), 157–160].
3. A note on Elliptic Functions [Quarterly Journal of Pure and Applied Mathematics, 21, (1886), 212–217].
4. On the differential equation of a trajectory [Journal of the Asiatic Society of Bengal, 56, Pt. II, No. 1, (1887), 117–120].
5. On Monge's differential equation to all conics [Journal of the Asiatic Society of Bengal, 56, Pt. II, No. 2, (1887), 134–145].
6. A memoir on plane analytical geometry [Journal of the Asiatic Society of Bengal, 56, Pt. II, No. 3, (1887), 288–349].
7. A general theorem on the differential equations of all trajectories [Journal of the Asiatic Society of Bengal, 57, Pt. II, No. 1, (1888), 72–99].
8. On Poisson's integral [Journal of the Asiatic Society of Bengal, 57, Pt. II, No. 1, (1888) 100–106].
9. On the differential equations of all parabolas [Journal of the Asiatic Society of Bengal, 57, Pt. II, No. 4, (1888), 316–332].

10. Remarks on Monge's differential equations of all conics [Proceedings of the Asiatic Society of Bengal, February, (1888)].

11. The geometric interpretation of Monge's differential equations of all conics [Journal of the Asiatic Society of Bengal, 58, Pt. II, (1889), 181–186].

12. Some applications of Elliptic Functions to problems of Mean Value (First paper) [Journal of the Asiatic Society of Bengal, 58, Pt. II, No. 2, (1889), 199–213].

13. Some applications of Elliptic Functions to problems of Mean Value (Second paper) [Journal of the Asiatic Society of Bengal, 58, Pt. II, No. 2, (1889), 213–231].

14. On Clebsch's transformation of the hydrokinetic equations [Journal of the Asiatic Society of Bengal, 59, Pt. II, No. 1, (1890), 56–59].

15. Note on Stoke's theorem and hydrokinetic circulation [Journal of the Asiatic Society of Bengal, 59, Pt. II, No. 1, (1890), 59–61].

16. On a curve of aberrancy [Journal of the Asiatic Society of Bengal, 59, Pt. II, No. 1, (1890), 61–63].

17. Mathematical Notes (Questions and Solutions) [Educational Times, London, 43, 44, 45, (1890–1892), 125–151, 144–182, 146–168].

List of Publications of Professor Syamadas Mukhopadhyay

1. 'Geometrical Theory of a Plane Non-Cyclic Arc, Finite as Well as Infinitesimal' [Journal of Asiatic Society of Bengal, (NS), IV, (1908)].

2. 'A General Theory of Osculating Conics I' [Journal and Proceedings of Asiatic Society of Bengal, (NS), IV, No. 4, (1908), 167–168].

3. 'A General Theory of Osculating Conics II' [Journal and Proceedings of Asiatic Society of Bengal, (NS), IV, No. 10, (1908), 497–509].

4. 'New Methods in the Geometry of a Plane Arc I, Cyclic and Sextactic Points' [Bulletin of Calcutta Mathematical Society, 1, (1909), 31–37].

5. 'On Rates of Variation of the Osculating Conic' [Bulletin of Calcutta Mathematical Society, No. 2, (1909), 125–130].

6. 'Parametric Coefficients in the Differential Geometry of Curves in an N-Space, I, General Conceptions' [Bulletin of Calcutta Mathematical Society, 1, No. 2, (1909), 187–200].

7. 'Parametric Coefficients in the Differential Geometry of Curves in an N-Space, II, Extension of Serret Frenet Formulae to Curves in an N-Dimensional Space' [Bulletin of Calcutta Mathematical Society, 1, No. 4, (1909), 233–234].

8. 'Parametric Coefficients in the Differential Geometry of Curves in an N-Space, III, Fundamental Formulae' [Bulletin of Calcutta Mathematical Society, 11, (1910)].

9. 'Parametric Coefficients in the Differential Geometry of Curves in an N-Space, IV, Expressions of the Coordinates of a Point on a Curve in an N-Space as Power Series in S' [Bulletin of Calcutta Mathematical Society, 111, (1911)].

10. 'Intrinsic Parameters in the Differential Geometry of Curves in an N-Space' [Griffiths Memorial Prize Essay, (1910), Calcutta University Publications].

11. 'Parametric Coefficients in the Differential Geometry of Curves, V, Principal Directions and Curvatures at a Singular Point of a Curve in an N-Space' [Bulletin of Calcutta Mathematical Society, V, (1913–14), 13–20].

12. 'A Note on the Stereoscopic Representation of Four-Dimensional Space' [Bulletin of Calcutta Mathematical Society, IV, (1912–13), 15].

13. 'Reply to Prof Bryan's Criticism' [Bulletin of Calcutta Mathematical Society, VI, (1914–15), 55–56].

14. 'Parametric Coefficients in the Differential Geometry of Curves in an N-Space, VI; On Parametric Coefficient and Osculating Spherics to a Curve in an N-Space' [Bulletin of Calcutta Mathematical Society, VIII, (1915)].

15. 'A Note on Current View of Operations Through the Fourth Dimension' [Bulletin of Calcutta Mathematical Society, IX, (1917)].

16. 'New Methods in the Geometry of a Plane Arc II Cyclic Points and Normals' [Bulletin of Calcutta Mathematical Society, X, (1919), 65–72].

17. 'Generalisations of Certain Theorems in the Hyperbolic Geometry of the Triangle' [Bulletin of Calcutta Mathematical Society, XII, (1920–21), 14–28].

18. 'Geometrical Investigations on the Correspondence Between a Right Angled Triangle, a Three Right Angled Quadrilateral and a Rectangular Pentagon in Hyperbolic Geometry' [Bulletin of Calcutta Mathematical Society, XIII, No. 4, (1922–23), 211–216].

19. 'Some General Theorems in the Geometry of Plane Curve' [Sir Astosh Mookerjee Silver Jubilee, II, (1922), Calcutta University Publications].

20. 'Genesis of an Elementary Arc' [Bulletin of Calcutta Mathematical Society, XVII, No. 4, (1926), 153].

21. With R C Bose 'On General Theorems of Co-intimacy of Symmetries and Hyperbolic Triad' [Bulletin of Calcutta Mathematical Society, XVII, No. 1, (1926), 39–55].

22. 'Triadic Equations in Hyperbolic Geometry' [Bulletin of Calcutta Mathematical Society, XVIII, No. 4, (1927)].

23. 'Note on T Hayashi's Paper on the Osculating Ellipses of a Plane Curve' [Circolo Mathematico Palermo, Tomo L 1, (1927)].

24. 'Generalied Form of Bohmer's Theorem for Elliptically Curved Non-Analytic Oval' [Mathematische Zeitschrift, Band 30, (1929), 560–571].

25. 'Extended Minimum Number Theorems of Cyclic and Sextactic on a Plane Convex Oval' [Mathematische Zeitschrift, Band 33, (1931), 648–662].

26. 'Circles Incident on an Oval of Undefined Curvature' [Tohuku Journal of Mathematics, Japan, 37, (1931)].

27. 'Lower Segments of M-Curves' [Journal of Indian Mathematical Society, XIX, (1931), 75–80].

28. 'Cyclic Curves of an Ellipsoid' [Journal of Indian Mathematical Society, XX, (1932), 246–250].

29. 'An Exposition of the Axioms of Order of Hilbert' [Indian Physico Mathematical Journal, III, (1932)].

30. 'Gemattic Extensions of Elementary Chains' [Tohuku Journal of Mathematics, Japan, 38, (1933)].

List of Publications of Professor Ganesh Prasad

1. On the Potential of Ellipsoids of Variable Densities [Messenger of Mathematics, 1901, 8].
2. Constitution of Matter and Analytical Theories of Heat [Abhandulgen d. k., Geseischalf der WissZu Gottingen, 1903].
3. Expansion of Arbitrary Functions in a Series of Spherical Harmonics [Math. Ann., 1912]. [This is a very important result and has been quoted in Hobson's book entitled 'Theory of Spherical and Ellipsoidal Harmonics, 148'].
4. On the Notions of Lines of Curvature [Proceedings of the Royal Society of Gottingen, 1904].
5. On the Present State of the Theory and Applications of Fourier's Series [Bulletin of the Calcutta Mathematical Society, 2, (1910–1911), 17–24].
6. On Some Recent Researches Relating to the Expansibility of Functions in Infinite Series [Bulletin of the Calcutta Mathematical Society, 2, (1–2), (1910–11), 3–9].
7. On the Existence of the Mean Differential Co-efficient of a Continuous Function [Bulletin of the Calcutta Mathematical Society, 3, (1911–1912), 53–54].
8. On a Non-Analytical Potential Function [Bulletin of the Calcutta Mathematical Society, 10, (1–4), (1908–1913), 39–41].
9. On the Foundations of the Theory of Surfaces [Bulletin of the Calcutta Mathematical Society, 10, (1–4), (1908–1913), 131–133].
10. On the Linear Distribution Corresponding to the Potential Function with a Prescribed Boundary Value [Bulletin of the Calcutta Mathematical Society, 5, (1913–1914), 47–52].
11. On the Second Derivates of the Newtonian Potential Due to a Volume Distribution Having a Discontinuity of the Second Kind [Bulletin of the Calcutta Mathematical Society, 6, (1914–1915), 47–52].
12. On the Vibrating String with an Infinite Number of Edges [Bulletin of the Calcutta Mathematical Society, 7, (1915–1916), 25–32].
13. On the Failure of Poisson's Equations and of Petrini's Generalization [Bulletin of the Calcutta Mathematical Society, 8, (1916–1917), 33–39].
14. On the Normal Derivative of the Newtonian Potential Due to a Surface Distribution Having a Discontinuity of the Second Kind [Bulletin of the Calcutta Mathematical Society, 9, (1917–1918), 1–9].
15. On the Fundamental Theorem of the Integral Calculus [Bulletin of the Calcutta Mathematical Society, 15, (1–4), (1924–1925), 57–68].
16. On the Fundamental Theorem of the Integral Calculus for Lebesgue Integrals [Bulletin of the Calcutta Mathematical Society, 16, (1926), 109–116].

17. On the Fundamental Theorem of the Integral Calculus in the Case of Repeated Integrals [Bulletin of the Calcutta Mathematical Society, 16, (1926), 1–8].
18. On the Summability (C1) of the Fourier Series of a Function at a Point Where the Function Has an Infinite Discontinuity of the Second Kind [Bulletin of the Calcutta Mathematical Society, 19, (1926), 51–58].
19. On the Failure of Lebesgue's Criterion for the Summability (C1) of the Fourier Series of a Function at a Point Where the Function Has a Discontinuity of the Second Kind [Bulletin of the Calcutta Mathematical Society, 19, (1926), 1–12].
20. On the Summability (C1) of the Derived Series of the Fourier Series of an Indefinite Integral Where the Integral Has a Discontinuity of the Second Kind [Bulletin of the Calcutta Mathematical Society, 19, (1926), 95–100].
21. On the Strong Summability (C1) of the Fourier Series of a Function at a Point Where the Function Has an Infinite Discontinuity of the Second Kind [Bulletin of the Calcutta Mathematical Society, 19, (1926), 127–134].
22. On the Failure of Lebesgue's Criterion for the Summability (C2) of the Fourier Series of a Function at a Point Where the Function Has a Certain Type of Discontinuity of the Second Kind [Bulletin of the Calcutta Mathematical Society, 19, (1926), 25–28].
23. On the Function θ in the Mean-Value Theorem of the Differential Calculus [Bulletin of the Calcutta Mathematical Society, 20, (1927), 155–184].
24. On the Differentiability of the Integral Function [Crelle's Journal, 160, (1929)]. [According to Mathematicians of the time it was an epoch making publication].
25. On Roll's Function as Multiple-Valued Function [Proceedings of the Benares Mathematical Society, X, 1929].
26. On the Summation of Infinite Series of Legendre's Functions [Bulletin of the Calcutta Mathematical Society, 22, (4), (1930), 159–170].
27. On the Zeros of Weierstrass's Non Differentiable Function [Proceedings of the Benares Mathematical Society, XI, (1930), 1–8].
28. On the Nature of θ in the Mean-Value Theorem of the Differential Calculus [Bulletin of the American Mathematical Society, 36, (1930)].
29. On the Summation of Infinite Series of Legendre's Functions (2nd Paper) [Bulletin of the Calcutta Mathematical Society, 23, (4), (1930), 115–124].
30. On the Determination of f(h) Corresponding to a Given Rolle's Function θ(h) When it is Multiple-Valued [Proceedings of the Benares Mathematical Society, XII, (1931)].
31. On Non-Orthogonal System of Legendre's Functions [Proceedings of the Benares Mathematical Society, XII, (1931)].
32. On the Differentiability of the Indefinite Integral and Certain Summability Criteria [Address Delivered in 1932 to Mathematical and Physical Section of Science Congress].
33. On the Lebesgue's Integral Mean-Value for a Function Having a Discontinuity of the Second Kind [Proceedings of the Benares Mathematical Society, XIV, (1933)].

34. On Lebesgue's Absolute Integral Mean-Value for a Function Having a Discontinuity of the Second Kind [Special memorial volume of the Tohoku Mathematical Journal in honour of Prof. Hayashi, (1933)].
35. 'Hobson, Presidential address on the life and works of the late Professor Hobson' [Bulletin of the Calcutta Mathematical Society, 25, (1933)].
36. Review: Lebesguesche Integrale und Fouriersche Reihen [Bulletin of the Calcutta Mathematical Society, XVII, (4), (1926), 203–206].

Books Written by Professor Ganesh Prasad

(1) Differential Calculus (1909).
(2) Integral Calculus (1910).
(3) An Introduction to Elliptic Functions (1928).
(4) A Treatise on Spherical Harmonics and the Functions of Bessel and Lamé (1930–32).
(5) Six Lectures on Recent Researches in the Theories of Fourier Series (1928).
(6) Six Lectures on Recent Researches About Mean-Value Theorem of the Differential Calculus.
(7) Mathematical Physics and Differential Equations at the Beginning of the 20th Century.
(8) Some Great Mathematicians of the 19th Century (2 Volumes).

List of Publications of Professor Bibhuti Bhusan Datta

Publications in Fluid Mechanics

1. On the method for determining the non-stationary state of heat in an ellipsoid [American Journal of Mathematics, 41, (1919), 133–142].
2. On the distribution of electricity in the two mutually influencing spheroidal conductor [Tohuku Mathematical Journal, Japan, (1920), 261–267].
3. On the stability of the rectilinear vortices of compressible fluids in an incompressible fluid [Philosophical Magazine, 40, (1920), 138–148].
4. Notes on vortices of a compressible fluid [Proceedings of the Benaras Mathematical Society, 2, (1920), 1–9].
5. On the stability of two co-axial rectilinear vortices of compressible fluids [Bulletin of the Calcutta Mathematical Society, 10 (4), (1920), 219–220].
6. On the periods of vibrations of straight vortex pair [Proceedings of the Benaras Mathematical Society, 3, (1921), 13–24].
7. On the motion of two spheroids in an infinite liquid along their common axis of revolution [American Journal of Mathematics, 43, (1921), 134–142].

Publications in History of Mathematics

1. Al-Biruni and the Origin of Arabic Numerals, Proceedings of the Benaras Mathematical Society, 7/8, pp 9–23, (1925–26).
2. A Note on Hindu–Arabic Numerals, American Mathematical Monthly, 33, pp 220–221, (1926).
3. Two Aryabhat as of Al-Birūni, Bulletin of the Calcutta Mathematical Society, 17, pp 59–74, (1926).
4. Hindu (Non-Jaina) Values of π, Journal of the Asiatic Society of Bengal, 22, pp 25–47, (1926c).
5. Early Literary Evidence of the Use of Zero in India, American Mathematical Monthly, 33, pp 449–454, (1926) and 38, p 566, (1931).
6. On Mūla, the Hindu Term for Root, American Mathematical Monthly, 34, pp 420–423, (1927a).
7. On the Origin and Development of the Idea of Percent, American Mathematical Monthly, 34, pp 530–531, (1927b).
8. Ārybhata, the Author of Ganita, Bulletin of the Calcutta Mathematical Society, 18, pp 5–18, (1927c).
9. Early History of the Arithmetic of Zero and Infinity in India, Bulletin of the Calcutta Mathematical Society, 18, pp 165–176, (1927d).
10. The Present Mode of Expressing Numbers, Indian Historical Quarterly, 3, pp 530–540, (1927e).
11. The Hindu Method of Testing Arithmetical Operations, Journal of the Asiatic Society of Bengal (n.s), 23, pp 261–267, (1927f).
12. Hindu Contributions in Mathematics, Bulletin of the Allahabad University Mathematical Association, 1, pp 49–72, (1927) and 36, (1929) (reprinted).
13. The Science of Calculation by the Board, American Mathematical Monthly, 35, pp 520–529, (1928a).
14. The Hindu Solution of the General Pellian Equation, Bulletin of the Calcutta Mathematical Society, 19, pp 87–94, (1928b).
15. On Mahavira's Solution of Rational Triangular and Quadrilaterals, Bulletin of the Calcutta Mathematical Society, 20, pp 267–294, (1928b).
16. SabdaSamkhyapranali (The Word Numeral System) (in Bengali), BangiyaSahityaParishadPatrika, Calcutta, for B.S 1335, pp 8–30, (1928–1929).
17. Vaidic O Poranic Shishumar SabdaSamkhyapranali (Vedic and Ancient Shishumar System of Writing Letter Numerals), (in Bengali), BangiyaSahitya-ParishadPatrika, Calcutta, for B.S 1335, pp 8–30, (1928).
18. The Bakhshali Mathematics, Bulletin of the Calcutta Mathematical Society, 21, pp 1–60, (1929).
19. The Jaina School of Mathematics, Bulletin of the Calcutta Mathematical Society, 21, pp 115–145, (1929).
20. The Scope and Development of Hindu Ganita, Indian Historical Quarterly, 5, pp 479–512, (1929).

21. A Short Review of G. R. Kaye, The Bakhshali Manuscript—A Study in Mediaeval Mathematics (Calcutta 1927), Bulletin of the CMS, 35, pp 579–580, (1929).

22. Aksara Samkhya Pranali (Alphabetical Numeral Systems) (in Bengali), BangiyaSahityaParishadPatrika, Calcutta for B.S 1336, pp 22–50, (1929–30).

23. Origin and History of the Hindu Names for Geometry, Quellen und Studien zur Geschichte der Mathematik BI, pp 113–119, (1930).

24. Geometry in Jaina Cosmography, Quellen und Studien zur Geschichte der Mathematik BI, pp 245–254, (1929–31).

25. On the Supposed Indebtedness of Brahmagupta to Chin–Chang Suan–Shu, Bulletin of the Calcutta Mathematical Society, 22, pp 39–51, (1930a).

26. The Two Bhaskaras, Indian Historical Quarterly, 6, pp 727–736, (1930).

27. On the Hindu Names for Rectilinear Geometrical Figures, Journal of the Asiatic Society of Bengal (n.s), 26, pp 283–290, (1930).

28. JyamitiShastrerprachin Hindu nam o taharprasar (Old Hindu Names of the Science of Geometry and Its Spread) (in Bengali), BangiyaSahityaParishad-Patrika, for B.S 1337 (1930), pp 1–6.

29. NamaSamkhya, (Nominal Numerals), (in Bengali), BangiyaSahityaParishad-Patrika, B.S 1337, pp 7–27 (1930–31a).

30. JainaSahityanamaSamkhya (Nominal Numerals in Jaina Literature) (in Bengali), BangiyaSahityaParishadPatrika, B.S 1337, pp 28–29, (1930–31).

31. AnkanamVamatogatih (in Bengali) BangiyaSahityaParishadPatrika, pp 70–80, (1930/31c).

32. Dashanka–Samkhyapranalirudbhavan (in Bengali) (Development of the System of Numerals by Ten Arithmetical Numbers), BangiyaSahityaParishadPatrika, Vol 3, 46th year.

33. On the Origin of the Hindu Term for 'Root', American Mathematical Monthly, 38, pp 371–376, (1931).

34. The Origin of Hindu Indeterminate Analysis, Archeion, 13, pp 401–407, (1931).

35. Narayana's Method for Finding Approximate Value of a Surd, Bulletin of the Calcutta Mathematical Society, 23, pp 187–194, (1931).

36. Early History of the Principle of Place Value, Scientica, 50, pp 1–2, (1931).

37. Early Literary Evidence of the Use of Zero in India (Second Article), American Mathematical Monthly, 38, (1931).

38. The Science of the Sulba, Calcutta University Publications, Calcutta, (1932).

39. Elder Aryabhatta's Rule for the Solution of Indeterminate Equations of the First Degree, Bulletin of the Calcutta Mathematical Society, 24, pp 19–36, (1932).

40. Testimony of Early Arab Writers on the Origin of Our Numerals, Bulletin of the Calcutta Mathematical Society, 24, pp 193–218, (1932).

41. On the Relation of Mahavira to Sridhara, ISIS 17, pp 25–33, (1932).

42. Introduction of Arabic and Persian Mathematics into Sanskrit Literature, Proceedings of the Benaras Mathematical Society, 14, pp 7–21, (1932).

43. 'Hindu GaniterAbanati' (Decline of Hindu Mathematics), (in Bengali), Panchapuspa B.S 1339, Month of Shravan (1932).

44. The Algebra of Narayana, ISIS 19, pp 427–485, (1933).

45. 'PrachinBangalijyotirbidMallikarjanSuri' (Mallikarjan Suri, the Old Bengali Astronomer), (in Bengali), BangiyaSahityaParishadPatrika for B.S 1340, No: 2, (1933).

46. Acharya Aryabhata and His Disciples and Followers, (in Bengali), BangiyaSahityaParishadPatrika for B.S 1340, pp 129–158, (1933/34).

47. MahabharateDashamka–Samkhya, (Numerals by Ten Arithmetical Numbers in Mahabharata) (in Bengali), BangiyaSahityaParishadPatrika, for B.S 1341, (1934).

48. Mathematics of Nemaichandra, JainaAntiquarry (Arrah), 1, pp 129 158, (1935).

49. Aryabhata and the Theory of the Motion of the Earth, (in Bengali), BangiyaSahityaParishadPatrika for B.S 1342, pp 167–183, (1935–36).

50. A Lost Jaina Treatise on Arithmetic, JainaAntiquarry (Arrah), 2, pp 38–40, (1936).

51. Vedic Mathematics, In cultural Heritage of India, III, pp 378–401, Calcutta, (1937).

52. Application of Intermediate Analysis to Astronomical Problems, Archeion, 21, pp 28–34, (1938–39).

53. Chronology of the History of Science in India During XVIth Century, Archeion, 23, pp 78–83.

54. Some Instruments of Ancient India and Their Working, Journal of the Ganganath Jha Research Institute, 4, pp 249–270, (BS 1337).

55. Hindu JyotiseShakKal, BangiyaSahityaParishadPatrika, B.S 1344, No: 3–4, pp 119–145.

56. BirshresthaArjunerBayas, BangiyaSahityaParishadPatrika, B.S 1344, No: 3–4, pp 186–200.

Books Written by Professor B. B. Datta

(1) The Sciences of Sulba, University of Calcutta, (1932).

(2) History of Hindu Mathematics, Part I (1930), Part II (1938), Part III published serially in Indian Journal of the History of Science of the Indian National Science Academy.

(3) Prachin Hindu Jyotisi, W.B State Book Board GanitCharcha, serially in 1983–89.

List of Publications of Professor Prasanta Chandra Mahalanobis

1. A new coefficient of correlation with applications to some biological and socio-logical data (Abstract) [Proceedings of the Indian Science Congress (Bombay), Section 6, (1919)] and [Proceedings of the Asiatic Society of Bengal, New Series, 15 (4); cxxiii, (1919)].

2. On the stability of anthropometric constants for Bengal caste data (Abstract) [Proceedings of the Indian Science Congress (Nagpur), Section Physics & Mathematics, 7, (1920)] and [Proceedings of the Asiatic Society of Bengal, New Series, 16, Iv–Ivi, (1920)].

3. (WITH SASANKA SEKHAR MUKHERJI) On the new compensated ballistic method for magnetic measurements, with a preliminary note on the magnetic behaviour of nickel in the form of powder under different physical stimuli (Abstract) [Proceedings of the Indian Science Congress (Nagpur), Section Physics & Mathematics, 7, (1920)] and [Proceedings of the Asiatic Society of Bengal, New Series, 16, iv, (1920)].

4. Note on the criterion that two samples are samples of the same population (Abstract) [Proceedings of the Indian Science Congress (Calcutta), 8, (1921)] and [Proceedings of the Asiatic Society of Bengal, New Series, 17, cxvii–cxviii, (1921)].

5. Anthropological observations on the Anglo-Indians of Calcutta, Part I: Head length and head breadth (Abstract) [Proceedings of the Indian Science Congress (Calcutta), 8, (1921)] and [Proceedings of the Asiatic Society of Bengal, New Series, 17, ccxIvii–ccxIviii, (1921)].

6. Anthropological observations on the Anglo-Indians of Calcutta, Part I: Analysis of meal stature [Records Indian Museum, 23, (1922), 1–96].

7. On the correction of a coefficient of correlation for observational errors (Abstract) [Proceedings of the Indian Science Congress (Madras), Section Physics and Mathematics, 9, (1922)] and [Proceedings of the Asiatic Society of Bengal, New Series, XVIII, (6), (1922), 54].

8. On the probable error of the component frequency constants of a dissected frequency curve (Abstract) [Proceedings of the Indian Science Congress (Madras), Section Physics and Mathematics, 9, (1922)] and [Proceedings of the Asiatic Society of Bengal, New Series, XVIII (6), (1922), 54].

9. On the probable error of constants obtained by linear interpolation (Abstract) [Proceedings of the Indian Science Congress (Madras), Section Physics and Mathematics, 9, (1922)] and [Proceedings of the Asiatic Society of Bengal, New Series, XVIII (6), (1922), 54].

10. On upper air correlations (Abstract) [Proceedings of the Indian Science Congress (Madras), Section Physics and Mathematics, 9, (1922)] and [Proceedings of the Asiatic Society of Bengal, New Series, XVIII (6), (1922), 53–54].

11. Correlation of upper air variables [Nature, 112, (1923), 323–324].

12. On errors of observation and upper air variables [Mem. India Met. Department, 24, (1923), 11–19].

13. On the seat of activity in the upper air [Mem. India Met. Department, 24, (1923), 1–9].

14. Statistical note on the significant character of local variation in the proportion of dextral and sinistral shells in samples of the Buliminus Dextro Sinister from the salt range, Punjab [Records Indian Museum, 25, (1923), 399–403].

15. On the probable error of interpolation for parabolic curves (Abstract) [Proceedings of the Indian Science Congress (Lucknow), Section Physics and Mathematics, 10, (1923), 77].

16. Statistical analysis of five independent samples of *cardina nilotica var gracilipes* (Abstract) [Proceedings of the Indian Science Congress (Lucknow), Section Physics and Mathematics, 10, (1923), 82].

17. New method of computing the rate of standard clocks (Abstract) [Proceedings of the Indian Science Congress (Bangalore), Section Mathematics and Physics, 11, (1924), 65].

18. Statistical studies in meteorology (Abstract) [Proceedings of the Indian Science Congress (Bangalore), Section Mathematics and Physics, 11, (1924), 47–48].

19. Analysis of race-mixture in Bengal [Presidential Address, Anthropological Section, Indian Science Congress (Edited): Journal Asiatic Society of Bengal, 23, (1925), 301–333].

20. Anthropometric survey of India (Abstract) [Proceedings of the Indian Science Congress (Bombay), Section Anthropology, 13, (1926), 320].

21. Appendicitis, rainfall and bowel complaints by Capt. S. K. Ray, Part II: Scope of the enquiry [Calcutta Medical Journal, 21 (4), (1926), 151–187, Discussion 213].

22. Correlation and variation of normal rainfall for July, August and September in North Bengal (Abstract) [Proceedings of the Indian Science Congress (Bombay), Section Mathematics and Physics, 13, (1926), 68].

23. Local variations of specimens of *cardina nilotica var gracilipes* (Abstract) [Proceedings of the Indian Science Congress (Bombay), Section Zoology, 13, (1926), 190].

24. Report on rainfall and flood in North Bengal during the period 1870–1922. (In 2 volumes with 29 tables and 28 maps) submitted to the Government of Bengal, 1–90 [Proceedings of the Indian Science Congress (Bombay), Section Mathematics and Physics, 14, (1927), (Abstract), Title: Rainfall in relation to floods in North Bengal].

25. On the need for standardization in measurements on the living [Biometrics, 20, A, (1928), 1–31] (Abstract) [Proceedings of the Indian Science Congress (Calcutta), Section Anthropology, 15, (1928), 325].

26. Statistical study of the Chinese head [Man in India, 8, (1928), 107–122]. [Proceedings of the Indian Science Congress (Calcutta), Section Anthropology, (Abstract), Title: 'A first study of Chinese head' 15, (1928), 325].

27. On tests and measures of group divergence. Part I: Theoretical formulae [Journal and Proceedings of the Asiatic Society of Bengal, (New Series), 26, (1930), 541–588].

28. Statistical study of certain anthropometric measurements from Sweden [Biometrics, 22, (1930), 94–108] (Abstract) [Proceedings of the Indian Science Congress (Allahabad), Section Anthropology, 17, (1930), 396].

29. Anthropological observations on the Anglo-Indians of Calcutta. Part II, Analysis of Anglo-Indian head length. [Records of Indian Museum, 21, (1930),

97–149] (Abstract) [Proceedings of the Indian Science Congress (Nagpur), Section Anthropology, 18, (1930), 411].

30. Applications of Statistics to agriculture [Proceedings of the Indian Science Congress (Nagpur), Section Agriculture, General Discussion, 18, (1931), 446–447].

31. On the normalization of statistical variates [Proceedings of the Indian Science Congress (Nagpur), Section Mathematics and Physics, 18, (1931), 91–92].

32. Revision of Risley's anthropometric data relating to the tribes and castes of Bengal (Abstract) [Proceedings of the Indian Science Congress (Nagpur), 18, (1931), 411], [Sankhyā, I, (1933), 76–105].

33. Statistical study of the level of the rivers of Orissa and the rainfall in the catchment areas during the period 1868–1928. Submitted to the Government of Bihar and Orissa.

34. Studies in group tests of intelligence: (1) The reliability and age-normals of scores in Form (A) (Abstract) [Proceedings of the Indian Science Congress (Nagpur), Section Psychology, 18, (1931), 437].

35. (WITH N. CHAKRAVARTI) Statistical report on river floods in Orissa during the period 1868–1928. (In 20 chapters with 47 tables and 27 maps), submitted to the Government of Bihar and Orissa.

36. Auxiliary tables for Fisher's Z-test in analysis of variance (Statistical notes for agricultural workers, no. 3) [Indian Journal of Agricultural Science, 2, (1932), 679–693].

37. Revision of Risley's anthropometric data relating to the Chittagong hill tribes (Abstract) [Proceedings of the Indian Science Congress (Bangalore), Section Anthropology, 19, (1932), 424], [Sankhyā, I, (1934), 267–276].

38. Rice and potato experiments at Sriniketan, (Agricultural Department of the Visva-Bharati), 1931. (Statistical notes for agricultural workers, no. 4). [Indian Journal of Agricultural Science, 2, (1932), 694–703] (Abstract) [Proceedings of the Indian Science Congress (Patna), Section Agriculture, 20, (1933), 48].

39. A statistical analysis of the height of the Brahmani River at Jenapore (Abstract) [Proceedings of the Indian Science Congress (Bangalore), Section Mathematics and Physics, 19, (1932), 123].

40. Statistical note on certain rice breeding experiments in the Central Provinces [Indian Journal of Agricultural Science, 2, (1932), 157–169] (Abstract) [Proceedings of the Indian Science Congress (Bangalore), Section Agricultural Statistical Methods, 19, (1932), 88].

41. Statistical note on the method of comparing mean values based on small samples [Indian Journal of Agricultural Science, 2, (1932), 28–41] (Abstract) [Proceedings of the Indian Science Congress (Bangalore), Section Agricultural Statistical Methods, 19, (1932), 88].

42. (WITH S. S. BOSE) A first study on sampling experiments on the effect of systematic arrangements in field trials (Abstract) [Proceedings of the Indian Science Congress (Bangalore), Section Agricultural Statistical Methods, 19, (1932), 88].

43. (WITH S. S. BOSE) Note on the variation of the percentage of infection of Wilt disease in the cotton (Statistical notes for agricultural workers, no. 5). [Indian Journal of Agricultural Science, 2, (1932), 704–709] (Abstract) [Proceedings of the Indian Science Congress (Patna), Section Agricultural Statistics, 20, (1933), 48].

44. A comparison of different statistical measures of intelligence based on a group test in Bengal (Abstract) [Proceedings of the Indian Science Congress (Patna), Section Psychology, 20, (1933), 439].

45. Editorial [Sankhyā, I, (1933), 1–4].

46. Effects of fertilizers on the variability of the yield and the rate of shedding of buds and flowers and bolls in the cotton plants in Surat (Statistical notes for agricultural workers, no. 6). [Indian Journal of Agricultural Science, 3, (1933), 131–138] (Abstract) [Proceedings of the Indian Science Congress (Patna), Section Agricultural Statistics, 20, (1933), 46].

47. A new photographic apparatus for recording profiles of living persons (Abstract) [Proceedings of the Indian Science Congress (Patna), Section Anthropology, 20, (1933), 413].

48. Notes on field technique. Submitted to the Department of Agriculture, Government of Bombay, (1933).

49. On the need of randomization of plots in field trials (Statistical notes for agricultural workers, no. 13) [Indian Journal of Agricultural Science, 3, (1933), 549–551] (Abstract) [Proceedings of the Indian Science Congress (Patna), Section Agricultural Statistics, 20, (1933), 48].

50. The reliability of a group test of intelligence in Bengali (with five figures), (Studies in educational test, No. 1) [Sankhyā, I, (1933), 25–49].

51. A study of the correlation between height of Brahmini at Jenapore and the rainfall in the catchment area (Abstract) [Proceedings of the Indian Science Congress (Patna), Section Mathematics and Physics, 20, (1933), 123].

52. Tables for the application of Neyman and Pearson's L-tests for judging the significance of divergence in mean values and variabilities of K samples (Abstract) [Proceedings of the Indian Science Congress (Patna), Section Mathematics and Physics, 20, (1933), 123].

53. Tables for comparing standard deviation of small samples (Abstract) [Proceedings of the Indian Science Congress (Patna), Section Mathematics and Physics, 20, (1933), 122].

54. Tables for L-tests [Sankhyā, I, (1933), 109–122].

55. Use of the method of paired differences for estimating the significance of field trials (Statistical notes for agricultural workers, no. 11) [Indian Journal of Agricultural Science, 3, (1933), 349–352] (Abstract) [Proceedings of the Indian Science Congress (Patna), Section Agricultural Statistics, 20, (1933), 41].

56. Use of random sampling numbers in agricultural experiments (Statistical notes for agricultural workers, no. 14) [Indian Journal of Agricultural Science, 3, (1933), 1108–1115].

57. (WITH K. C. BASAK) A study of the intensity of floods in the Mahanadi for the period 1868–1929 (Abstract) [Proceedings of the Indian Science Congress (Patna), Section Mathematics and Physics, 20, (1933), 124].

58. (WITH S. S. BOSE) Analysis of a manorial experiment on wheat conducted at Sakrand, Sind (Statistical notes for agricultural workers, no. 10) [Indian Journal of Agricultural Science, 3, (1933), 345–348].

59. (WITH S. S. BOSE) Analysis of vertical tests with wheat conducted at Sakrand, Sind (Statistical notes for agricultural workers, no. 12) [Indian Journal of Agricultural Science, 3, (1933), 544–548] (Abstract) [Proceedings of the Indian Science Congress (Patna), Section Agricultural Statistics, 28, (1933), 47].

60. (WITH S. S. BOSE) Certain varietal studies on the cotton plant in Surat (Statistical notes for agricultural workers, no. 9) [Indian Journal of Agricultural Science, 3, (1933), 339–344].

61. (WITH S. S. BOSE) Effects of different doses of nitrogen on the rate of shedding of buds, flowers and bolls in the cotton plant in Surat (Statistical notes for agricultural workers, no. 8) [Indian Journal of Agricultural Science, 3, (1933), 147–154] (Abstract) [Proceedings of the Indian Science Congress (Patna), Section Agricultural Statistics, 28, (1933), 47].

62. (WITH S. S. BOSE) A study of the intensity of floods in the Brahmini for the period 1868–1929 (Abstract) [Proceedings of the Indian Science Congress (Patna), Section Mathematics and Physics, 20, (1933), 124].

63. (WITH NISTARAN CHAKRAVARTI) A study of the precipitation and the percentage run-off in the Mahanadi (Abstract) [Proceedings of the Indian Science Congress (Patna), Section Mathematics and Physics, 20, (1933), 124].

64. (WITH KEDARNATH DAS) A preliminary note on the rates of maternal deaths and still-births in Calcutta [Sankhyā, I, (1933), 215–230].

65. (WITH A. C. NAG) A study on the areal distribution of rainfall during rain storms in the Mahanadi catchment (Abstract) [Proceedings of the Indian Science Congress (Patna), Section Mathematics and Physics, 20, (1933), 123].

66. (WITH P. R. RAY) A study of the correlation of the height of Mahanadi at Sambalpur and at Naraj (Abstract) [Proceedings of the Indian Science Congress (Patna), Section Mathematics and Physics, 20, (1933), 124].

67. (WITH R. N. SEN) A study of the seasonal fluctuations in the height of the Orissa rivers (Abstract) [Proceedings of the Indian Science Congress (Patna), Section Mathematics and Physics, 20, (1933), 124].

68. Age variation of scores in a group test of intelligence in Bengali (with eight figures) (Studies in educational tests, No. 2) [Sankhyā, I, (1934), 231–244].

69. A statistical analysis of rotational experiments with cotton, groundnut and juar in Berar, with notes on designs of rotational experiments (Statistical notes for agricultural workers, no. 15) [Indian Journal of Agricultural Science, 4, (1934), 361–385] (Abstract) [Proceedings of the Indian Science Congress (Bombay), Section Agricultural Statistics, 21, (1934), 69].

70. [Note by Editor appended to] 'Tables for testing the significance of linear regression in the case of time-series and other single-valued samples by S. S. Bose' [Sankhyā, I, (1934), 284].

71. [Note by Editor appended to] 'Thirty-eight years of rice-yields in lower Birbhum, Bengal by Hashim Amir Ali, assisted by Tara Krishna Bose' [Sankhyā, I, (1934), 387–389].

72. On the statistical divergence between certain species of phytophthora (Abstract) [Proceedings of the Indian Science Congress (Bombay), Section Mathematics and Physics, 21, (1934), 151].

73. A preliminary note on intervarietal correlation in the rice plant in Bengal (Abstract) [Proceedings of the Indian Science Congress (Bombay), Section Agricultural Statistics, 21, (1934), 70].

74. A preliminary study of the intelligence quotient of Bengali school children (with ten figures) (Studies in educational tests, No. 4) [Sankhyā, I, (1934), 407–426].

75. (WITH S. S. BOSE) Effects of different types of soil-covers on the moisture economy of irrigated plantations of *Dalbergia Sisoo* in Lahore Division (Abstract) [Proceedings of the Indian Science Congress (Bombay), Section Agricultural Statistics, 21, (1934), 70].

76. (WITH S. S. BOSE) Statistical note on the effect of pests on the yield of sugarcane and the quality of cane-juice [Sankhyā, I, (1934), 399–496] (Abstract) [Proceedings of the Indian Science Congress (Bombay), Section Agricultural Statistics, 21, (1934), 71].

77. (WITH S. S. BOSE et al.) Tables for random samples from a normal population [Sankhyā, I, (1934), 289–328].

78. (WITH K. N. CHAKRAVARTI) Analysis of marks in the school leaving certificate examination in the United Provinces, India, 1919 (with ten figures) (Studies in educational tests, No. 3) [Sankhyā, I, (1934), 245–266].

79. (WITH S. C. CHAKRAVARTI and E. A. R. BANERJEE) Influence of shape and size of plots on the accuracy of field experiments with rice, Chinsurah, Bengal (Abstract) [Proceedings of the Indian Science Congress (Bombay), Section Agricultural Statistics, 21, (1934), 71].

80. (WITH KEDARNATH DAS and WITH ANIL CHANDRA NAG) A preliminary note on the rates of maternal deaths and still-births in Calcutta [Sankhyā, I, (1934), 215–230].

81. (WITH ANIL CHANDRA NAG) A preliminary study of the maternal death-rate and proportion of still-births in Bengal for the period 1848–1901 (Abstract) [Proceedings of the Indian Science Congress (Bombay), Section Medical and Veterinary Research, 21, (1934), 380].

82. Analysis of racial likeness in Bengal castes (Abstract) [Proceedings of the Indian Science Congress (Calcutta), Section Anthropology, 22, (1935), 355].

83. Application of statistical method in industry [Science and Culture, 1, (1935), 73–78].

84. [An Editorial correction to] 'On the distribution of ratio variances of two samples drawn from a given normal bivariate correlated population by Bose, S. S.' [Sankhyā, 2, (1935), 72].

85. Further studies of the Bengali profile (Abstract) [Proceedings of the Indian Science Congress (Calcutta), Section Anthropology, 22, (1935), 337].

86. On the validity of a group test of intelligence in Bengal (Abstract) [Proceedings of the Indian Science Congress (Calcutta), Section Psychology, 22, (1935), 446].

87. A statistical note on certain haematological studies of fifty new-born (Abstract) [Proceedings of the Indian Science Congress (Calcutta), Section Medical and Veterinary Research, 22, (1935), 411].

88. Study of consumer preference in Calcutta. Submitted to D. J. Keymer & Co.

89. (WITH K. C. BANERJEE and P. R. RAY) Studies in tiller-formation (Abstract) [Proceedings of the Indian Science Congress (Calcutta), Section Agricultural Statistics, 22, (1935), 347].

90. (WITH S. S. BOSE) Extension of X-table (corresponding to Fisher's Z-table) for testing the significance of two observed variances (Abstract) [Proceedings of the Indian Science Congress (Calcutta), Section Mathematics and Physics, 22, (1935), 79].

91. (WITH R. C. BOSE) On generalized statistical distance between samples from two normal populations (Abstract) [Proceedings of the Indian Science Congress (Calcutta), Section Mathematics and Physics, 22, (1935), 80].

92. (WITH S. S. BOSE) A note on the application of multiple correlation for estimating the individual digestibilities of a mixed feed (Abstract) [Proceedings of the Indian Science Congress (Calcutta), Section Agricultural Statistics, 22, (1935), 348].

93. (WITH S. S. BOSE) On estimating individual yields in the case of mixed—up—yields of two or more plots in agricultural experiments [Science and Culture, I, (1935), 205].

94. (WITH S. S. BOSE and S. C. CHAKRAVARTI) A complex cultural experiment with rice (Abstract) [Proceedings of the Indian Science Congress (Calcutta), Section Agricultural Statistics, 22, (1935), 347].

95. (WITH S. S. BOSE and S. C. CHAKRAVARTI) Complex experiment on rice at the Chinsurah farm, Bengal, 1933–1934 (Statistical notes for agricultural workers, no. 16) [Indian Journal of Agricultural Science, 6, (1935), 34–51].

96. (WITH S. S. BOSE and C. J. HARRISON) Effect of manurial dressings, weather conditions and manufacturing processes on the quality of tea at Tocklai experimental station, Assam [Sankhyā, 2, (1935), 33–42].

97. (WITH S. S. BOSE and T. V. G. MENON) A statistical study under permanent manurials in Pusa (Abstract) [Proceedings of the Indian Science Congress (Calcutta), Section Agricultural Statistics, 22, (1935), 348].

98. (WITH S. S. BOSE and S. RAY CHOUDHURY) A bivariate sampling experiment (Abstract) [Proceedings of the Indian Science Congress (Calcutta), Section Mathematics and Physics, 22, (1935), 80].

99. (WITH P. C. DAS and N. K. RAY CHOUDHURY) A statistical analysis of hospital records of some delivery cases in Calcutta (Abstract) [Proceedings of the Indian Science Congress (Calcutta), Section Medical and Veterinary Research, 22, (1935), 410].

100. (WITH T. V. G. MENON and S. S. BOSE) A statistical study of soil deterioration in the permanent manorial experiments in Pusa (Abstract) [Proceedings

of the Indian Science Congress (Calcutta), Section Agricultural Statistics, 22, (1935), 348].

101. (WITH G. C. NANDI et al.) Relation between heights and weights of Bengali women (Abstract) [Proceedings of the Indian Science Congress (Calcutta), Section Medical and Veterinary Research, 22, (1935), 410].

102. (WITH K. C. RAY) A statistical method of testing the genuineness of a sample of ghee (Abstract) [Proceedings of the Indian Science Congress (Calcutta), Section Medical and Veterinary Research, 22, (1935), 410].

103. Appendix I, Editorial note on the fundamental formula [to]: 'The yield of Andropogon Sorghum in relation to circumference of head, length of head and height of plant by Venkataramanen, S. N.' [Sankhyā, 2, (1936), 263–272].

104. Editorial note on the margin of error in the calculation of the cost of cultivation and profit Appended to 'Marketing of rice at Bolpur, by Satya Priya Bose' [Sankhyā, 2, (1936), 121–124].

105. Karl Pearson (1857–1936) [Lecture delivered at Statistical Laboratory, Calcutta, 27.11.1936] [Sankhyā, 2, (1936), 303–378].

106. New theory of ancient Indian chronology [Sankhyā, 2, (1936), 309–320].

107. (WITH S. ROY) School marks and intelligence test scores [Sankhyā, 2, (1936), 397–402].

108. Note on the statistical and biometric writings of Karl Pearson (with references) [Sankhyā, 2, (1936), 411–422].

109. Note on cotton prices in relation to quality and yield [Sankhyā, 2, (1936), 135–142].

110. Note on the use of indices in anthropometric work [Science and Culture, I, (1936), 477] (Abstract) [Proceedings of the Indian Science Congress (Indore), Section Anthropology, 23, (1936), 392].

111. On the generalised distance in statistics [Proceedings of the National Institute of Sciences, India, 11 (1), (1936), 49–55].

112. On the generalized measure of divergence between statistical groups (Abstract) [Proceedings of the Indian Science Congress (Indore), Section Mathematics and Physics, 23, (1936), 108].

113. Sir Rajendra Nath Mookerjee: First President of the Indian Statistical Institute (1931–1936) [Sankhyā, 2, (1936), 237–240].

114. (WITH D. P. ACHARYA) A statistical study of marks in the annual and test examinations in relation to University results in I. A. and I. Sc. examinations in Bengal (Abstract) [Proceedings of the Indian Science Congress (Indore), Section Psychology, 23, 536].

115. (WITH K. C. BANERJEE and S. S. BOSE) The influence of the date of planting and the number of seedlings per hole on tillering in rice at Bankura (Abstract) [Proceedings of the Indian Science Congress (Indore), Section Agriculture, 23, (1936), 437].

116. (WITH R. C. BOSE and S. N. ROY) On the evaluation of the probability integral of the D^2 statistic (Abstract) [Proceedings of the Indian Science Congress (Indore), Section Mathematics and Physics, 23, (1936), 107].

117. (WITH S. S. BOSE) On the estimate of missing yields in a splitplot type of arrangement (Abstract) [Proceedings of the Indian Science Congress (Indore), Section Agriculture, 23, (1936), 426].

118. (WITH S. S. BOSE) A situation experiment with rice (Abstract) [Proceedings of the Indian Science Congress (Indore), Section Agriculture, 23, (1936), 426].

119. (WITH S. S. BOSE and K. C. BANERJEE) Studies in tiller variation (Statistical notes for agricultural workers, no. 20) [Indian Journal of Agricultural Science, 6, (1936), 1122–1133].

120. (WITH S. S. BOSE and S. C. CHAKRAVARTI) Effect of different methods of harvest on the estimated error of field experiments on rice (Statistical notes for agricultural workers, no. 21) [Journal of Agricultural Live-stock, India, 6, (1936), 814–825].

121. (WITH S. S. BOSE and P. M. GANGULY) Frequency distribution of plot yields and optimum size of plots in a uniformity trial with rice in Assam (Statistical notes for agricultural workers, no. 19) [Indian Journal of Agricultural Science, 6, (1936), 1107–1121].

122. (WITH S. S. BOSE and R. K. KULKARNI) On the influence of shape and size of plots on the effective precision of field experiments with juar (Andropogon Sorghum) (Statistical notes for agricultural workers, no. 17) [Indian Journal of Agricultural Science, 6, (1936), 460–474].

123. (WITH S. S. BOSE and S. C. SENGUPTA) Statistical analysis of a manurial experiment on Napier grass (Pennisetum perpureum) by the method of covariance (Statistical notes for agricultural workers, no. 18) [Journal of Agricultural Live-stock, India, 6, (1936), 460–474].

124. (WITH S. C. CHAKRAVARTI and S. S. BOSE) A complex cultural experiment with rice at Chinsurah, Bengal. For the year 1934–35 (Abstract) [Proceedings of the Indian Science Congress (Indore), Section Agriculture, 23, (1936), 436].

125. (WITH K. K. GUHA ROY) Statistical methods and their applications to agronomy—a bibliography [Misc. Bul. No. 9, Coun. Agric. Res. India], 120 pages.

126. (WITH J. C. GUPTA) A preliminary note on measurements of blood pressure (Abstract) [Proceedings of the Indian Science Congress (Indore), Section Medical and Veterinary Research, 23, (1936), 485].

127. (WITH S. ROY) School marks and intelligence test scores [Sankhyā, 2, (1936), 377–402].

128. (WITH J. C. SEN) A comparative study of measures of intelligence (Abstract) [Proceedings of the Indian Science Congress (Indore), Section Psychology, 23, (1936), 536].

129. Need of a sample survey of the growth of population in India [Sankhyā, 3, (1937), 58].

130. A note on forecasting value of intelligence test (Abstract) [Proceedings of the Indian Science Congress (Hyderabad), Section Psychology, 24, (1937), 448].

131. Variation of rainfall with lunar periods in Calcutta for the month of July (Abstract) [Proceedings of the Indian Science Congress (Hyderabad), Section

Mathematics and Physics, 24, (1937), 72] (altered title) 'Note on the influence of lunar phase on the rainfall in the month of July in Calcutta 1878–1924' [Sankhyā, 3, (1937), 233–238].

132. On the accuracy of profile measurements with a photographic profiloscope [Sankhyā, 3, (1937), 65–72] [Slightly altered title]: 'Studies with the photographic profiloscope' (Abstract) [Proceedings of the Indian Science Congress (Indore), Section Anthropology, 23, (1936), 39].

133. Rectangular co-ordinates in sampling distributions: Appendix (Note) [Sankhyā, 3, (1937), 35].

134. Review of the application of the statistical theory to agricultural field experiments in India [Proceedings of the Second Meeting Crops and Soils Wing, Board of Agriculture and Animal Husbandry, Lahore, 6 December, (1937), Government of India, 1–14]. Revised paper [Indian Journal of Agricultural Science, 10, (1937), 192–212].

135. Statistical note on the Hooghly-Howrah flushing irrigation scheme. Submitted to the Government of Bengal, Irrigation Department, (1937), 1–44.

136. (WITH R. C. BOSE and S. N. ROY) The use of intrinsic rectangular co-ordinates in the theory of distribution (Abstract) [Proceedings of the Indian Science Congress (Hyderabad), Section Mathematics and Physics, 24, (1937), 99].

137. (WITH R. C. BOSE and S. N. ROY) Normalization of statistical variates and the rectangular co-ordinates in the theory of distribution [Sankhyā, 3, (1937), 1–34, Appendix by P. C. Mahalanobis 35–40].

138. (WITH D. P. ACHARYA) Note on the correlation between results in the College and University examinations [Sankhyā, 3, (1937), 239–244].

139. (WITH S. S. BOSE) On an exact test of association between the occurrence of thunderstorm and an abnormal ionization [Sankhyā, 3, (1937), 249–252] [Slightly altered title]: 'On a method of testing the association between thunderstorm and upper air ionization' (Abstract) [Proceedings of the Indian Science Congress (Hyderabad), Section Mathematics and Physics, 24, (1937), 99].

140. (WITH B. N. DATTA) Note on the foot and stature correlation of certain Bengali Castes and tribes [Sankhyā, 3, (1937), 245–248].

141. First session of the Indian Statistical Conference, Calcutta [Sankhyā, 4, (1938), 1–4].

142. Note on grid sampling [Science and Culture, 4, (1938), 300].

143. On the distribution of Fisher's taxonomic coefficient (Abstract) [Proceedings of the Indian Science Congress (Calcutta), Section Mathematics and Physics, 25, (1938), 31].

144. On an improved model of the profiloscope (Abstract) [Proceedings of the Indian Science Congress (Calcutta), Section Anthropology, 25, 206].

145. Professor Ronald Aylmer Fisher [Sankhyā, 4, (1938), 265–272] [Biometrics (with few minor editorial amendments), 20, (1938), 238–251].

146. Report on the sample census of jute in 1938 [Submitted to the Indian Central Jute Committee].

147. Statistical report on the experimental crop census, 1937 [Submitted to the Indian Central Jute Committee (1938)].
148. (WITH K. C. BANERJEE and J. R. PAL) Study on tillers of rice plant bearing on their duration on life and performance and death [Sankhyā, 4, (1938), 149] (Abstract) [Proceedings of the Indian Science Congress (Calcutta), Section Agriculture, (1938), 25, part III, Section IX, 220].
149. (WITH S. S. BOSE) Test of significance of treatment means with mixed up yields in field experiments (Abstract) [Proceedings of the Indian Science Congress (Calcutta), Section Agriculture, (1938), 25, part III, Section IX, 219].
150. (WITH S. C. CHAKRAVARTI and S. S. BOSE) Complex cultural experiment of rice (Abstract) [Sankhyā, 4, (1938), 149].
151. (WITH S. HEDAYATULLAH and K. P. ROY) Complex experiment of winter rice at Dacca (1936–37) [Sankhyā, 4, (1938), 149–150] (Abstract) [Proceedings of the Indian Science Congress (Calcutta), Section Agriculture, (1938), 25, part III, Section IX, 220].
152. Enquiry into the prevalence of drinking tea among middle class Indian families in Calcutta, 1939 [Submitted to the Tea Market Expansion Board, 1939] First Report on the Crop Census of 1938 [Indian Central Jute Committee, (1939), 1–110].
153. A note on grid sampling (Abstract) [Proceedings of the Indian Science Congress (Lahore), Section Mathematics and Physics, 26, Part III, Sec. I, 7].
154. Progress report of the jute census scheme for 1939 [Submitted to the Indian Central Jute Committee (1939)].
155. Review of the application of statistical theory to agricultural field experiments in India [Proceedings of the Second Meeting Crops Soils Wing, Agriculture, India, (1939), 200–215].
156. Subhendu Sekhar Bose, (1906–1932) [Sankhyā, 4, (1939), 313–336].
157. The technique of random sample survey (Abstract) [Proceedings of the Indian Science Congress (Lahore), Section Theoretical Statistics, 26, Part IV, Section II, 14–16].
158. (WITH S. S. BOSE and K. L. KHANNA) Note on the optimum shape and size of plots for sugarcane experiments in Bihar (Statistical notes for agricultural workers, no. 24) [Indian Journal of Agricultural Science, 9, (1939), 807–816].
159. (WITH S. C. CHAKRAVARTI and S. S. BOSE) Complex cultural experiment on rice (Abstract) [Proceedings of the Indian Science Congress (Calcutta), Section Agriculture, (1938), 25, III, 220].
160. (WITH K. R. NAIR and S. C. CHAKRAVARTI) A 10×10 quasifactorial experiment at Chinsurah with 100 strains of rice (Abstract) [Proceedings of the Indian Science Congress (Lahore), Section Agriculture, (1939), 26, Part III, Section IX, 199].
161. Applications of statistical methods in physical anthropometry [Proceedings of the Second Indian Statistical Conference, Lahore. Sankhyā, 5, (1940), 594–598] Altered Title: 'Application of statistical methods in anthropological

research' (Abstract) [Proceedings of the Indian Science Congress (Lahore), Section Anthropology, (1939), 26, Part IV, Section VIII, 23].

162. Characteristic features of rain storms and river floods in Orissa [Sankhyā, 5, (1940), 601–602].

163. Discussion on planning of experiments [Sankhyā, 5, (1940), 530–531].

164. Errors of observation in physical measurements [Science and Culture, 5, (1940), 443–445].

165. Note on the expenditure on tea among working class families in Howrah and Kankinara [Submitted to the Tea Market Expansion Board, 1940].

166. Preliminary report on the sample census of the area under jute [Submitted to the Indian Central Jute Committee (1940)].

167. Rain storms and river floods in Orissa (with a map) [Sankhyā, 5, (1940), 1–20].

168. Report on the Sample Census of Jute in 1939 [Submitted to the Indian Central Jute Committee (1940), 1–146].

169. Statistical note on crop-cutting experiments on paddy in Mymensingh [Submitted to the Government of Bengal, (1940)].

170. Statistical note on crop-cutting experiments on jute in 1940 [Submitted to the Indian Central Jute Committee (1940)].

171. (WITH K. R. NAIR) Simplified method of analysis of quasifactorial experiments in square lattice with a preliminary note on joint analysis of yield of paddy and straw (Statistical notes for agricultural workers, no. 25) [Indian Journal of Agricultural Science, 10, (1940), 663–685].

172. (WITH S. SEN) Fertility rates based on sample survey [Sankhyā, 5, (1940), 60].

173. General report on the sample census of the area under jute in Bengal, 1941 [Submitted to the Indian Central Jute Committee (1941), 43].

174. Note on random fields [Science and Culture, 7, (1941), 54].

175. On non-normal fields (Abstract) [Proceedings of the Indian Science Congress (Banaras), Mathematics and Statistics, 28, (1941), Section I, Part III, 12].

176. On the sample survey of jute in Bengal [Indian Central Jute Committee, Calcutta, 1938–1941].

177. Preliminary statistical report on the regional survey of Borer pests of sugarcane. December, 1940–March, 1941 [Submitted to the Imperial Council of Agricultural Research, June, 1941].

178. Report on programme preference and broadcast reactions, Calcutta, April–May, 1941 [Submitted to the Department of Information, Government of India, 1941].

179. Report on the sampling technique for forecasting the bark-yield of Cinchona plants: Experiment series B, 1940–1941 [Submitted to the Superintendent, Cinchona Culture in Bengal, Mungpoo, October, 1941].

180. Statistical note on nutritional investigations in college hostels in Calcutta [Sankhyā, 5, (1941), 439–448].

181. Statistical report on crop-cutting experiments on jute in 1940 [Submitted to the Indian Central Jute Committee, June, 1941].

182. Statistical report on the rupee census—Mathematical Appendix (Revised) [Report Currency and Finance, 1940–41, Reserve Bank of India, 49–55].

183. Statistical report on a sample survey of the prevalence of drinking tea in Nagpur City in 1940 [Submitted to the Tea Market Expansion Board, March, 1941].

184. Statistical survey of public opinion [Modern Review, 69, (1941), 393–397].

185. (WITH C. BOSE) Correlation between anthropometric characters in some Bengal castes and tribes [Sankhyā, 5, (1941), 249–269].

186. (WITH K. R. NAIR) Statistical analysis of experiments on different Limen values for lifted weights [Sankhyā, 5, (1941), 285–294].

187. Family budget enquiries of labourers, Jagaddal, Scheme for Labour Family Budget Enquiry and Preparation of cost of Living Index of Industrial Workers [Submitted to the Board of Economic Enquiry, Bengal, January, 1942].

188. Note on the life of a rupee note, July, 1940–June, 1941 [Reserve Bank of India, 1942].

189. Preliminary report on Burdwan–Hooghly–Howrah crop-cutting survey [Submitted to the Government of Bengal, Irrigation Department, (1942)].

190. Preliminary report on the crop-cutting experiments on jute in 1942 [Submitted to the Government of Bengal, 1942].

191. Preliminary report on the sample census of the area under jute in Bengal, 1942 [Submitted to the Government of Bengal, 1942].

192. Report on the precision of family budget enquiry at Jagaddal, August, 1942 [Submitted to the Board of Economic Enquiry, Bengal, 1942].

193. Report on the sample survey of jute and aus paddy crops.

194. Sample survey of public opinion (Abstract) [Proceedings of the Indian Science Congress, (Baroda), Section Mathematics and Statistics, 29, (1942), Part III, Sec I, 15].

195. Sample surveys: Presidential Address, Section Mathematics and Statistics, Indian Science Congress, (Baroda), (1942) (Abstract) [Proceedings of the Indian Science Congress, (Baroda), 29, (1942), Part II, 25–46] [Science and Culture, 7, No. 10 (Supplement), (1942), April 1–2].

196. (WITH C. BOSE) On devising an efficient sampling technique for forecasting the mean value of a variable (Abstract) [Proceedings of the Indian Science Congress, (Baroda), Section Mathematics and Statistics, 29, (1942), Part III, Sec I, 14].

197. (WITH K. GUPTA) Enquiry into the family budgets of labourers in Bengal (Abstract) [Proceedings of the Indian Science Congress, (Baroda), Section Mathematics and Statistics, 29, (1942), Part III, Sec I, 15].

198. (WITH N. T. MATHEW) On the rupee census problem (Abstract) [Proceedings of the Indian Science Congress, (Baroda), Section Mathematics and Statistics, 29, (1942), Part III, Sec I, 14–15].

199. Enquiry into the prevalence of drinking tea among middle class Indian families in Calcutta, 1939; 'Studies in Sample Surveys' [Sankhyā, 6, (1943), 283–312].

200. Final report on the sample census of the area under jute and aus paddy in Bengal 1943—Sample surveys for estimating the acreage and yield of jute

in Bengal 1937–1943 [Submitted to the Government of Bengal, 1–51, (37 Tables)].

201. Preliminary report on the sample census of the area under aman paddy in Bengal 1943 [Submitted to the Government of Bengal (1943)].

202. Report on the regional survey of Borer pests of sugarcane. December, 1940–March, 1941 [Submitted to the Imperial Council of Agricultural Research, (1943), 1–284].

203. (WITH B. N. GHOSH) Statistical analysis of data relating to incidence of pests and diseases on different varieties of sugarcane [Proceedings of the Indian Statistical Conference, Calcutta, 1943] (Abstract) [Sankhyā, 4, (1939), 349].

204. [Editorial Note] to 'The stability of the income distribution by Harro Bernardelli' [Sankhyā, 6, (1944), 362].

205. Multi-stage sampling (Abstract) [Proceedings of the Indian Science Congress, (Delhi), Section Mathematics and Statistics, 31, Part III, Section I, 3–4].

206. On large-scale sample surveys [Philosophical Transactions of the Royal Society, London, Series B, 231, (1944), 329–451].

207. Organisation of Statistics in the post-war period [Proceedings of the National Institute of Sciences, India, 10, (1944), 69–78].

208. Preliminary report on the crop-cutting experiments on aus paddy, 1942 [Submitted to the Government of Bengal (1944)].

209. Report on Bengal crop survey 1943–1944. Part I—Jute and aus paddy 1943. Part 2—Aman paddy, 1943–1944, 46 pages.

210. Bengal crop survey 1944–1945: Progressive reports on bhadoi (monsoon) crops, Submitted on 8 and 20 October and 4 November, 1944; on aghani (winter) crops on 11 November, 1, 8 and 25 December, 1944, 2 and 22 January, 2 and 9 February and 21 March, 1945; and on robi (spring) crops on 19 and 23 March 1945.

211. Bengal crop survey 1944–1945: Progressive reports. Submitted to the Government of Bengal on jute and aus paddy crops on 31 August, 16 September and 14 October 1944 and on aman paddy crops on 19 December 1944 and 16 and 24 December 1944 and 16 and 24 January and 1, 8, 14 and 23 February, 1945.

212. Note on the mortality in Bengal in 1943 [Submitted to the Famine Inquiry Commission in February, 1945].

213. Note on the number of destitute in 1943 [Submitted to the Government of Bengal on 28 May 1945].

214. Report on the Bihar crop survey, Rabi season, 1943–44 [Submitted to the Government of Bihar] [Sankhyā, 7, (1945), 29–106].

215. Bengal famine: the background and basic facts [Paper read at East-Asia Association at the Royal Society 25 July, 1945] [Asiatic Review, October (1946), 7].

216. A direct method of estimating total production of crops (Abstract) [Proceedings of the Indian Science Congress, (Bangalore), Section Statistics, 33, Part III, Section II, (1946), 17].

217. Distribution of the Muslims in population of India [Sankhyā, 7, (1946), 429–434].

218. Problems of current demographic data in India [In] 'Papers presented by special guests of the Population Association of America. October 25–26, 1946' New York (mimeo).

219. Recent experiments in statistical sampling in the Indian Statistical Institute [Asia Publishing House and Statistical Publishing Society, 1961, 9, 70 pages].

220. Sample surveys on crop yields in India [Sankhyā, 7, (1946), 269–280].

221. Traffic census on new Howrah bridge [Submitted to the Government of India, (1946)].

222. Use of small size plots in sample surveys for crop yields [Nature, 158, (1946), 798–799].

223. (WITH D. N. MAJUMDAR and C. R. RAO) Biometric analysis of anthropological measurements on castes and tribes of the United Provinces (Abstract) [Proceedings of the Indian Science Congress, (Bangalore), Section Anthropology and Archaeology, 33, Part III, Section 8, 138].

224. (WITH R. K. MUKHERJI and A. GHOSE) Sample survey of after-effects of the Bengal famine of 1943 [Sankhyā, 7, (1946), 337–400].

225. 'Famine and Rehabilitation in Bengal' [Calcutta Statistical Publishing Society, (1946), 1–63].

226. Enquiry into the economics of agricultural labour and rural indebtedness [Submitted to the Government of India, (1947)].

227. Enquiry into the economics and statistics of road developments. Reports of the traffic and economic surveys conducted at Sherghatti, Jehanabad, Bariarpur, Mohania and Toposi [Submitted to the Government of India, (1947)].

228. On the combination of data from test conducted at different laboratories. Summary of lecture reported by Tucker, J. (Jr.) [American Society for Testing Materials (ASTM) Bulletin, 144, (1947), 64–66].

229. Report on tour of Canada, U. S. A. and U. K. October 15 to December 15, 1946 [Sankhyā, 8, (1947), 403–410].

230. Walter A. Shewhart and statistical quality control in India [Sankhyā, 9, (1949), 51–60].

231. Historical note on D^2—statistic, Appendix I, Anthropological Survey of United Provinces, 1941: a statistical study [Sankhyā, 9, (1949), 237–239].

232. Statistical tools in resource appraisal and utilization. UN, (1949), 1–14. [U. N. Scientific Conference on the conversion and utilization of resources, 17 August–6 September, 1949, Lake Success, New York, Vol. 1, (Plenary Meetings), (1949), 196–200].

233. United Nations. Economic and Social Council. Sub-commission on statistical sampling (held from 30 August to 11 September 1948) {Chairman: P. C. Mahalanobis} [Sankhyā, 9, (1949), 377–398].

234. (WITH D. N. MAJUMDAR and C. R. RAO) Anthropometric survey of United Provinces, 1941: a statistical study [Sankhyā, 9, (1949), 90–324].

235. Age tables based on the Y-samples [Submitted to the Government of India, (1950)].

236. Cost and accuracy results in sampling and complete enumeration [Bulletin of International Statistical Institute, 32 (2), (1950), 210–213].

237. Survey of economic condition of agricultural labour (1946–47) [Submitted to the Government of West Bengal, (1950)].

238. Survey of rural indebtedness. Final Report: Rural indebtedness Enquiry, 1946–47 [Submitted to the Government of West Bengal, (1950)].

239. Syllabus for an advanced (Professional) course in Statistical Sampling [Sankhyā, 10, (1950), 152–154]. {Part of Appendix 'A' of UNESCO. Sub-commission on statistical sampling. Report on the 3rd session [Sankhyā, 10, (1950), 129–158]}.

240. Why Statistics? General Presidential Address, Indian Science Congress, Thirty-seventh Session, Poona (2nd January, 1950) [Proceedings of the Indian Science Congress, (Poona), 37, Part II, (1950), 1–32] [Sankhyā, 10, (1950), 195–228].

241. In memorium: Abraham Wald [Sankhyā, 12, (1951), 1–2].

242. India: Ministry of Finance, Ministry of Economic Affairs—National Income Committee. Chairman: P. C. Mahalanobis. First report—1951, 102 pages.

243. Professional training in statistics [Bulletin International Statistical Institute, 33 (5), (1951), 335–342].

244. Means of livelihood and industries tables based on Y-sample [Submitted to the Government of India, (1951)].

245. Role of mathematical statistics in secondary education [Bulletin International Statistical Institute, 33 (5), (1951), 323–334].

246. (WITH J. M. SENGUPTA) On the size of sample cuts in crop-cutting experiments in the Indian Statistical Institute, 1939–1950. Appendices A & B [Bulletin International Statistical Institute, 33 (3), (1951), 359–404].

247. National income, investment and national development. Lecture delivered at the National Institute of Sciences of India, New Delhi, 4 October 1952 [Talks on planning. Asia Publishing House and Statistical Publishing Society, 1961, 9–12].

248. Some aspects of the design of sample surveys [Sankhyā, 13, (1952), 1–7].

249. Statistical methods in national development. The Thirteenth 'Jagadish Chandra Bose Memorial Lecture', Bose Institute, Calcutta, 1951 [Science and Culture, 17, (1952), 497–504].

250. National Sample Survey: General Report No. 1 on the first round, October 1950–March 1951 [Submitted to the Government of India, (1952)] [Sankhyā, 13, (1952), 47–214].

251. Some observations on the process of growth of national income [Sankhyā, 13, (1952), 307–312].

252. (WITH S. B. SEN) On some aspects of the Indian national sample survey [Bulletin International Statistical Institute, 34 (2), (1953), 5–14].

253. Foundations of statistics [Dialectica, 8, (1954), 95–111] [Sankhyā, 18, (1957), 183–194].

254. India: Ministry of Finance, Ministry of Economic Affairs—National Income Committee. Final report. Chairman: P. C. Mahalanobis, 1954, 173 pages.

255. Report on survey of saver's preference in Delhi state [Submitted to the Ministry of Finance, Government of India, (1954)].

256. Studies relating to planning for national development. Address delivered on the occasion of inauguration by Prime Minister Jawaharlal Nehru of the 'Studies relating to planning for national development', at the Indian Statistical Institute, Calcutta, 3 November, 1954 [Talks on planning. Asia Publishing House and Statistical Publishing Society, 1961, 13–18].

257. Approach to planning in India. Based on a talk broadcast from All India Radio, 11 September, 1955 [Talks on planning. Asia Publishing House and Statistical Publishing Society, 1961, 47–54].

258. Approach of operational research to planning in India [Sankhyā, 16, (1955), 3–62] [An Approach of Operational Research to Planning in India; Asia Publishing House and Statistical Publishing Society, 1961, vi, 168p].

259. Draft plan-frame for the Second Five Year Plan 1956–1961 [Science and Culture, 20, (1955), 619–632].

260. Draft plan-frame for the Second Five Year Plan. 1956/57–1960/61: recommendations for the formulation for Second Five Year Plan [Submitted to the Government of India, 17 March, (1955)] [Sankhyā, 16, (1955), 63–90] [Published as 'recommendations for the ... plan' [in] Talks on planning. Asia Publishing House and Statistical Publishing Society, 1961, 19–46].

261. Agricultural statistics in relation to planning. Address delivered at the Ninth Annual Meeting of the Indian Society of Agricultural Statistics, 7 January, 1956 [Journal of Indian Society of Agricultural Statistics, 8, (1956), 5–13].

262. The Geological, Mining and Metallurgical Society of India, Calcutta. Address delivered at the Thirty Second Annual General Meeting of 28 September, 1956 [Quarterly Journal of Geology, Mining and Metallurgical Society, India, 28, (1956), 87–88].

263. Statistics must have purpose. Presidential Address, Pakistan Statistical Conference, Lahore, February, 1956.

264. Some impressions of a visit to China, 17 June–11 July, 1957. Typed pages 35.

265. Statistics, a survey, (University teaching of social science, No. 7). Prepared and edited on behalf of the International Statistical Institute, The Hague, Paris, UNESCO, 1957, 209 pages.

266. Enfoque de la planeaction en la India [Trim. Econ., 25, (1958), 654–663].

267. Industrialization of underdeveloped countries—a means to peace. Paper presented at the Third Pugwash Conference at Kitzbuhel-Vienna, September, 1958 [Bulletin Atom. Scient., 15 (1), (1959), 12–17] [Sankhyā, 22, (1960), 173–180] [Talks on planning. Asia Publishing House and Statistical Publishing Society, 1961, 125–136 (with appendix)].

268. Industrializatsiya—klyuch kukrepleniyu nezavisimosti. (Industrialization—a key to the consolidation of independence), [Sovremennyi Vostok (Contemporary East), No. 12, (1958), 15–18].

269. Methods of fractile graphical analysis with some surmises of results [Transactions of Bose Research Institute, 22, (1958), 223–230].

270. Relaxation of tensions through Industrialization of the underdeveloped countries (Mimeo-graph) Kitzbuhel-Vienna Conference of Scientists, September, 1958.

271. Science and national planning. Anniversary Address delivered at the National Institute of Science of India, Madras, 5 January, 1958 [Sankhyā, 20, (1958), 69–106] [Talks on planning. Asia Publishing House and Statistical Publishing Society, 1961, 55–92] [Science and Culture, 23, (1958), 396–410].

272. Some observations on the 1960 world census on agriculture [Bulletin International Statistical Institute, 36 (4), (1958), 214–221].

273. Heralding a new epoch [in] 'A Study of Nehru' [Ed. Rafiq Zakaria, (1959), 309–320] [Talks on planning. Asia Publishing House and Statistical Publishing Society, 1961, 1–8].

274. Izuchenie problem industrializatii slaborazvitikh stran (Study of problems of industrialization in the underdeveloped countries). Sovremennyi Vostok (1959) (Contemporary East) [No. 9, English translation in Talks on planning. Asia Publishing House and Statistical Publishing Society, 1961, 137–142].

275. Need of scientific and technical man-power for economic development. Based on a talk broadcast from All India Radio, 23 September, 1959 [Talks on planning. Asia Publishing House and Statistical Publishing Society, 1961, 143–146].

276. Next steps in planning. Anniversary Address delivered at the National Institute of Science of India, New Delhi, 20 January, 1959 [Talks on planning. Asia Publishing House and Statistical Publishing Society, 1961, 1–8] [Sankhyā, 22, (1960), 143–172].

277. Problems of economic development in India and other underdeveloped countries in relation to world affairs, [Bulletin of the International House of Japan, 3, (1959) (10–15)].

278. Review of recent developments in the organization of science in India. Presented at the 6th General Assembly and Scientific symposium held under the auspices of the World Federation of Scientific Workers, 1959 [Vijnan Karmee, 11, (1959), 13–27].

279. Unemployment and underemployment. Address delivered as the Sectional Chairman of the Second All India Labour Economics Conference, Agra, January, 1959 [Indian Journal of Labour Economics, 2, (1959), 39–45] [Talks on planning. Asia Publishing House and Statistical Publishing Society, 1961, 147–152].

280. (WITH A. DAS GUPTA) The use of sample surveys in demographic studies in India [UN World Population Conference, Rome, (1954)] [E/Conf. 13/418, VI, (1959), 363–384].

281. Economic development of Afro-Asian countries (Prepared for Bandung Conference, April, 1955). Appendix to 'Industrialization of underdeveloped countries—a means to peace' by P. C. Mahalanobis [Sankhyā, 22, (1960), 181–182].

282. Incentives and scientific and technical personnel [Dainik Samachar, (1960)].

283. Labour problems in mixed economy. Presidential address delivered at the Third All India Labour Economic Conference, Madras, 2 January, 1960 [Indian Journal of Labour Economics, 3, (1960), 1–8] [Talks on planning. Asia Publishing House and Statistical Publishing Society, 1961, 153–159].

284. Methods of fractile graphical analysis [Econometrica, 28, (1960), 325–351] [Sankhyā, 23 A, (1961), 41–64].
285. Note on problems of scientific personnel. Draft recommendations placed before the Scientific Personnel Committee, 24 March, 1960 [Science and Culture, 27, (1960), 40–128].
286. On the use of fractile graphical method for analysis of economic data (Abstract) [Proceedings of the Indian Science Congress, (Bombay), Section Statistics, 47, Part III, Section II, (1960), 23].
287. Perspective planning. Address delivered at the Third Session of the SEANZA Central Banking Course in Bombay (1960).
288. Scientific workers in the U. K., U. S. A. and the U. S. S. R. [Science and Culture, 27, (1960), 101–110].
289. (WITH D. B. LAHIRI) Analysis of errors in censuses and surveys with special reference to experience in India [Bulletin International Statistical Institute, 38 (2), (1960), 409–433] [Sankhyā, 23 A, (1961), 325–358].
290. Preliminary note on the consumption of cereals in India. (With 4 appendices) [Bulletin International Statistical Institute, 39 (4), (1960), 53–76] [Sankhyā, 25 B, (1963), 217–236].
291. (WITH M. MUKHERJEE) Operational research models used for planning in India. Presented at the Second Operational Research International Congress, September, 1960.
292. Role of science in economic and national development. Lecture delivered at the University of Sofia, 4 December, 1961 [Indian Journal of Public Administration, 8, (1961), 153–160].
293. Statistics for economic development [Journal of the Royal Society of Japan, 3, (1961), 97–112] [Sankhyā, 27 B, (1965), 179–188].
294. 'Talks on planning' (collected addresses and broadcasts on various aspects of planning). Asia Publishing House and Statistical Publishing Society (1961) [Studies relating to Planning for National Development, No. 6] [Indian Statistical Series No. 14], iv, 159 pages.
295. Scientific base on economic development. Presented at the Conference for 'International Cooperation and Partnership', Salzburg-Vienna, 1–7 July, 1962 [Sankhyā, 25 B, (1963), 55–56].
296. First Convocation of the Indian Statistical Institute—Section II Review of the Director by P. C. Mahalanobis [Science and Culture, 28, (1963), 92–97].
297. Introducing volume twenty-five [Sankhyā, 25 A, (1963), 1–4 and 25 B, (1963), Corrigenda 427].
298. Need of a standard terminology for classification of different types of research [Science and Culture, 29, (1963), 224–225].
299. Recent developments in the organization of science in India. [Sankhyā, 25 B, (1963), 67–84, Corrigenda 426].
300. Social transformation for national development [Sankhyā, 25 B, (1963), 49–57, Corrigenda 426].
301. Some personal memories of R. A. Fisher [Sankhyā, 25 A, (1963), 1–4] [Biometrics, 20, (1964), 368–371].

302. (WITH R. K. SOM and H. MUKHERJEE) Analysis of variance of demographic variables. (Note by Editor, P. C. Mahalanobis) [Sankhyā, 24 B, (1963), 21–22].

303. Problems of internal transformation. Presented at the Conference for 'International Cooperation and Partnership', Salzburg-Vienna, 1–7 July, 1962.

304. Statistical tools and techniques in perspective planning in India [Bulletin International Statistical Institute, 40 (1), (1963), 152–169] [Sankhyā, 26 B, (1964), 29–44].

305. India, Planning Commission Report of the committee on distribution of income and levels of living. Chairman: P. C. Mahalanobis, Part I: Distribution of Income and Wealth and concentration of economic power, (1964), 107 pages.

306. Objects of science education in underdeveloped countries. (The Commonwealth Conference on teaching of science in schools, Colombo, December 1963) [Sankhyā, 26 B, (1964), 253–256] [Slightly altered title] 'The aims of science teaching in schools—science education in underdeveloped countries' [Commonwealth Education Committee: School science teaching—report of an expert conference held at the University of Ceylon, Paradeniya. Appendix II (ii), (1964), 28–31].

307. Perspective planning in India: statistical tools [Coexistence, (1964), May, 60–73].

308. Priorities in science in underdeveloped countries [The twelfth Pugwash Conference, Udaipur, January 27–February 1, (1964), 181–193] [Sankhyā, 26 B, (1964), 45–52].

309. Some concepts of sample surveys in demographic investigations. Presented at the U. N. World Population Conference, Belgrade, August–September, 1965, Vol. III, 246–250. [Reprinted with changes] [Sankhyā, 28 B, (1966), 199–204].

310. Statistics as a key technology [Annals of Statistics, 19 (2), (1965), 43–46].

311. Use of capital output ratio in planning in developing countries [Proceedings of the 35th Session, International Statistical Institute, Belgrade, September, 1965] [Bulletin International Statistical Institute, 41 (1), (1965), 87–95] [Sankhyā, 29 B, (1967), 249–256].

312. Extensions of fractile graphical analysis to higher dimensional data. RTS Technical Report No. 7/66, February 1966, 1–13 (mimeographed). [Essays in Probability and Statistics (S. N. Roy Memorial Volume). Calcutta Statistical Publishing Society, 1969, 397–406].

313. Objectives of science and technology. Presented before the Symposium on Collaboration between the Countries of Africa and Asia for the Promotion and Utilization of Science and Technology, New Delhi, April–May, 1966. [Seminar, 82, (1966), 38–43].

314. Quality control for economic growth [Sankhyā, 29 B, (1967)], [Inaugural Address, 4th All India Conference on Statistical Quality Control, Madras, 7–9 December, 1967]. Technical Report No. 41/69.

315. Royal Society Conference on Commonwealth Scientist [Science and Culture, 33, (1967), 149–153].

316. The Asian Drama: an Indian review [Sankhyā, 31 B, (1969), 435–458] [Shorter Review in Scientific American, 22 (1), 1969 July, 128–134] [Longer version, Economic Policies Weekly, 4 (28), (1969), 30: Special Number, July 1969; 1119–1132, July 1969].

317. Basic problems of design of sample surveys [International Conference on Computer Science organized by the Institute of Statistical Studies and Research, Cairo University, December, 1969].

318. Extensions of fractile graphical analysis [Proceedings of International Conference on Quality Control, Tokyo, ICQC, 1969, 515–518].

319. (WITH D. P. BHATTACHARYYA) Growth of population in India and Pakistan: 1800–1961 [General Congress, 1969, International Union for Scientific Study of Population, London. Technical Report No. Demo/6/69].

320. India. Planning Commission. Report of the Committee on distribution of income and levels of living [Chairman: P. C. Mahalanobis] Part II: Changes in levels of living, 1969, 114 pages.

321. (WITH R. C. BOSE et al.) (Eds.) Essays in probability and statistics. Calcutta Statistical Publishing Society, 1969.

322. Social change, science and economic growth. One Asia Assembly on New Directions for Asia, organised by Press Foundation of Asia, Manila, Philippines, April 1970.

323. (WITH D. B. LAHIRI et al.) Technical aspects of design, National Sample Survey. RTS publications: mimeograph series, 226 pages.

324. Some observations on recent developments in sample surveys [Proceedings of the 38th Session (Washington), Bulletin International Statistical Institute, 44 (1), (1971), 247–261. Discussion 262–268].

List of Publications of Professor Nikhil Ranjan Sen

1. 'On the potentials of heterogeneous incomplete ellipsoids and elliptic discs.' [Bulletin of the Calcutta Mathematical Society, 10 (1918) 157].

2. 'On the external potential of infinite elliptic cylinders.' [Philosophical Magazine, 38 (1919) 465].

3. 'On a type of vibration of a thin elastic spherical shell in a gaseous medium.' [Philosophical Magazine, 42 (1921) 192].

4. 'The equation of long waves in canals of varying section.' [Philosophical Magazine, 48 (1924) 65].

5. 'Note on the propagation of waves in elastic media.' [Bulletin of the Calcutta Mathematical Society, 16 (1924) 9].

6. 'On the boundary conditions for the gravitational field equations on surfaces of discontinuity.' [Annls Phys, 4 (1924) 73].

7. 'On de Sitter's Universe' (with M V Laue) [Annls Phys, 4 (1924) 74].

8. 'On the calculation of the fall of potential in the ions and electron gas in contact with glowing metals.' [Annls Phys, 4 (1924) 82].

9. 'On the electric particle according to Einstein's field theory.' [Zeitschrift fur Physik, 40 (1927) 667].

10. 'On Fresnel's convection in general relativity.' [Proceedings of the Royal Society, 116 (1927) 73].

11. 'On the separation of H-lines in parallel and crossed electric and magnetic fields.' [Zeitschrift fur Physik, 65 (1929) 673].

12. 'Equations of the electron theory and Dirac's wave mechanics (I).' [Zeitschrift fur Physik, 66 (1930) 122].

13. 'On the Kepler problem for the five dimensional generalised wave equation and the influence of gravitational field on spectral lines.' [Zeitschrift fur Physik, 66 (1930) 686].

14. 'Motion of material particles in a homogeneous gravitational field.' [Zeitschrift fur Physik, 66 (1930) 693].

15. 'Equations of the electron theory and Dirac's wave mechanics (II).' [Zeitschrift fur Physik, 68 (1931) 267].

16. 'On the interpretation of Dirac's matrices.' [Indian Phys. Math. Journal, 2 (1931) 1].

17. 'Radiation in expanding universe.' [Indian Phys. Math. Journal, 3 (1932) 89].

18. 'On Edington's problem of the expansion of the universe by condensation.' [Proceedings of the Royal Society, 140 (1933) 269].

19. 'On Schwarzschild's problem of the gaseous sphere.' [Zeitschrift fur Astro-Physik, 7 (1933) 188].

20. 'On the equilibrium of an incompressible sphere' (with N K Chatterjee) [Monthly Notices of the Royal Astronomical Society, 94 (1934) 550].

21. 'On the stability of cosmological model.' [Zeitschrift fur Astro-Physik, 9 (1934) 215].

22. 'Minimum property of the Friedman Space.' [Zeitschrift fur Astro-Physik, 9 (1935) 315].

23. 'Stability of the cosmological models.' [Zeitschrift fur Astro-Physik, 10 (1935) 29].

24. 'Rate of disappearance of the proper motion of a Nebula according to the expansion theory.' [Bulletin of the Calcutta Mathematical Society, 27 (1935) 101].

25. 'Principle of equivalence and deduction of Lorentz transformation.' [Indian Journal of Physics, 10 (1936) 341].

26. 'Size of dense spheres.' [Zeitschrift fur Astro-Physik, 14 (1937) 157].

27. 'Expansion of a Nebula.' [Bulletin of the Calcutta Mathematical Society, 29 (1937) 185].

28. 'Two elementary theorems on polytropes.' [Bulletin of the Calcutta Mathematical Society, 30 (1938) 11].

29. 'Pressure relations in the interior of Stellar bodies.' [Zeitschrift fur Astro-Physik, 18 (1939) 124].

30. 'On theoretical estimates of an upper limit of Stellar diameters.' [Indian Journal of Physics, 15 (1941) 209].

31. 'On some thermodynamical properties of a mixture of gas and radiation.' [Indian Journal of Physics, 15 (1941) 219].
32. 'On the inversion of density gradient and convection in Stellar bodies.' [Proceedings of the National Institute of Science, India, 7 (1942) 183].
33. 'On Stellar models based on Bethe's law of energy generation.' [Proceedings of the National Institute of Science, India, 8 (1942) 317].
34. 'Contribution to the theory of Stellar models.' [Proceedings of the National Institute of Science, India, 8 (1942) 339].
35. 'A note on the meson wave.' [Bulletin of the Calcutta Mathematical Society, 34 (1942) 61].
36. 'An approximate solar model based on Bethe's law of energy generation' (with U R Burman) [Astrophysics Journal, 100 (1944) 247].
37. 'Note on the Cowling model of a convective radiative star' (with U R Burman) [Indian Journal of Physics, 18 (1944) 212].
38. 'On the internal constitution of stars of small masses according to Bethe's law of energy generation' (with U R Burman) [Astrophysics Journal, 102 (1945) 208].
39. 'Note on large scale motion in viscous stars' (with N L Ghosh) [Bulletin of the Calcutta Mathematical Society, 37 (1945) 141].
40. 'The problem of internal constitution of stars.' [Bulletin of the Calcutta Mathematical Society, 38 (1946) 1].

List of Publications of Professor Suddhodan Ghosh

1. On liquid motion inside certain rotating circular arc [Bulletin of the Calcutta Mathematical Society, 15, (1924), 27–46].
2. On a problem of elastic circular plates [Bulletin of the Calcutta Mathematical Society, 16, (1925), 63–70].
3. On the solution of $\Delta^4_1 w = C$ in bipolar coordinates and its application to a problem in elasticity [Bulletin of the Calcutta Mathematical Society, 16, (1925), 117–122].
4. On certain many valued solutions of the equations of elastic equilibrium and their application to the problem of dislocation in bodies with circular boundaries [Bulletin of the Calcutta Mathematical Society, 17, (1926), 185–194].
5. On the steady motion of viscous liquid due to translation of a tore parallel to its axis [Bulletin of the Calcutta Mathematical Society, 18, (1927), 185].
6. On plane strain and stress in rotating elliptic cylinders and discs [Bulletin of the Calcutta Mathematical Society, 19, (1928), 117–126].
7. On the bending of a loaded elliptic plate [Bulletin of the Calcutta Mathematical Society, 21, (1929), 191–194].
8. On the solution of the equations of elastic equilibrium suitable for elliptic boundaries [Transactions of the American Mathematical Society, 32, (1930), 47].

9. On the stress and strain in a rolling wheel [Bulletin of the Calcutta Mathematical Society, 25, (1933), 99–106].

10. Flexure of beams of certain forms of cross-sections [Bulletin of the Calcutta Mathematical Society, 27, (1935), 61–68].

11. A note on the vibrations of a circular ring [Bulletin of the Calcutta Mathematical Society, 27, (1935), 177–182].

12. Plane strain in an infinite plate with an elliptic hole [Bulletin of the Calcutta Mathematical Society, 28, (1936), 21–47].

13. On some simple distributions of stress in three dimensions [Bulletin of the Calcutta Mathematical Society, 28, (1936), 107–119].

14. Stress distribution in a heavy circular disc held with its plane vertical by a peg at the centre [Bulletin of the Calcutta Mathematical Society, 28, (1936), 145–150].

15. On the solutions of Laplace's equation suitable for problems relating to two spheres touching each other [Bulletin of the Calcutta Mathematical Society, 28, (1936), 193–198].

16. On some two dimensional problems of elasticity [Bulletin of the Calcutta Mathematical Society, 28, (1936), 213–222].

17. On the distribution of stress in a semi-infinite plate under the action of a couple at a point in it [Bulletin of the Calcutta Mathematical Society, 29, (1937), 177–184].

18. Stress distribution in an infinite plate containing two equal circular holes [Bulletin of the Calcutta Mathematical Society, 31, (1939), 149–159].

19. On plane strain and plane stress in aeolotropic bodies [Bulletin of the Calcutta Mathematical Society, 34, (1942), 157–169].

20. Stress systems in rotating aeolotropic discs [Bulletin of the Calcutta Mathematical Society, 35, (1943), 61–65].

21. On the divergence of the solution of a problem of plane strain [Bulletin of the Calcutta Mathematical Society, 36, (1944), 51–58].

22. A note on average stresses in a plate [Bulletin of the Calcutta Mathematical Society, 38, (1946), 10–20].

23. On the concept of generalized plane stress [Bulletin of the Calcutta Mathematical Society, 38, (1946), 45–56].

24. On generalized plane stress in an aeolotropic plate [Bulletin of the Calcutta Mathematical Society, 38, (1946), 61–66].

25. On the flexure of an isotropic elastic cylinder [Bulletin of the Calcutta Mathematical Society, 39, (1947), 1–14].

26. On a new function—theoretic method of solving the torsion problem for some boundaries [Bulletin of the Calcutta Mathematical Society, 39, (1947), 107–112].

27. On the flexure of a beam whose cross-section is bounded partly by a straight line [Bulletin of the Calcutta Mathematical Society, 40, (1948), 77–82].

28. On the torsion and flexure of a beam whose cross-section is a quadrant of a given area [Bulletin of the Calcutta Mathematical Society, 40, (1948), 107–115].

29. Torsion of a solid of revolution of a material possessing curvilinear aeolotropy [Journal of the Association of Applied Physics, Calcutta University, 3, (1956), 1–4].

List of Publications of Professor Rabindranath Sen

1. Simplexes in n-dimensions [Bulletin of Calcutta Mathematical Society, 18, (1926), 33–64].
2. Infinitesimal analysis of an arc in n-space [Proceedings of Edinburgh Mathematical Society, (1928), 149–159].
3. Spherical simplexes in n-dimensions [Proceedings of Edinburgh Mathematical Society, Ser. 2, Part I (1930), 1–10].
4. On the new field theory [Indian Physico Mathematical Journal, No. 2, (1930), 28–31].
5. On curvatures of a hypersurface [Bulletin of Calcutta Mathematical Society, 23, (1931), 1–10].
6. On rotations in hypersurfaces [Bulletin of Calcutta Mathematical Society, 23, (1931), 195–209].
7. On the connection between Levi-Civita parallelism and Einstein's teleparallelism [Proceedings of Edinburgh Mathematical Society, Ser. 2, Part 4 (1931), 252–255].
8. Note on tubes of electromagnetic forces [Bulletin of Calcutta Mathematical Society, 35, (1933), 191–196].
9. On a type of three dimensional space compatible with Clifford's parallelism [Tohuku Journal of Mathematics, 42, Part 2, (1936), 226–229].
10. Parallelism in Riemannian space [Bulletin of Calcutta Mathematical Society, 36, (1944), 102–107].
11. Parallelism in Riemannian space II [Bulletin of Calcutta Mathematical Society, 37, (1945), 153–159].
12. Parallelism in Riemannian space III [Bulletin of Calcutta Mathematical Society, 38, (1946), 161–167].
13. Parallel displacement and scalar product of vectors [Proceedings of the National Institute of Sciences, India, 14, (1948), 45–52].
14. Parallel displacement and scalar product of vectors II [Bulletin of the Calcutta Mathematical Society, 41, (1949), 41–46].
15. Parallel displacement and scalar product of vectors III [Bulletin of the Calcutta Mathematical Society, 41, (1949), 113–120].
16. On an algebraic system generated by a single element and its application in Riemannian Geometry [Bulletin of the Calcutta Mathematical Society, 42, (1950), 1–13].
17. On an algebraic system generated by a single element and its application in Riemannian Geometry II [Bulletin of the Calcutta Mathematical Society, 42, (1950), 117–187].

18. On an algebraic system generated by a single element and its application in Riemannian Geometry III [Bulletin of the Calcutta Mathematical Society, 43, (1951), 77–94].

19. Corrections to my papers on an algebraic systems etc. [Bulletin of the Calcutta Mathematical Society, 44, No. 2, (1952)].

20. On a type of vector spaces [Proceedings of the National Institute of Sciences, India, 19, (1953), 475–486].

21. On pairs of teleparallelism [Journal of the Indian Mathematical Society, 17, (1953), 21–32].

22. On pairs of teleparallelism II [Journal of the Indian Mathematical Society, 19, (1955), 61–71].

23. Note on non-simple K^*-space [Proceedings of the National Institute of Sciences, India, 22, (1956), 82–85].

24. Parallelism in Differential Geometry [Presidential address at the 43rd Session of the Indian Science Congress, Agra, (1956)].

25. On a type of Riemannian space conformal to a flat space [Journal of the Indian Mathematical Society, 21, (1958), 105–114].

26. A note on symmetric affine connections [Indian Journal of Mathematics, 1, (1958), 17–19].

27. On a geometry at a point of a hypersurface of a Riemannian space [Bulletin of the Calcutta Mathematical Society, 50, (1958), 193–203].

28. (With H. Sen): On a generalization of a space of constant curvature [Bulletin of the Calcutta Mathematical Society, Golden Jubilee Commemoration Volume, (1958), 129–139].

29. On a correspondence between a system of second-order symmetric tensors and a system of affine connections [Proceedings of the National Institute of Sciences, India, 26, A, Suppl. II, (1960), 14–20].

30. On a sequence of conformal Riemannian spaces [Bulletin of the Calcutta Mathematical Society, 54, (1962), 107–121].

31. On an algebraic system of conformal Riemannian spaces [Indian Journal of Mathematics, 4, (1962), 71–85].

32. Associate tensor and affine connections [Journal of the Indian Mathematical Society, 27, (1963), 45–56].

33. On new theories of space in general relativity [Bulletin of the Calcutta Mathematical Society, 56, (1964), 1–14].

34. On new theories of space in unified field theory [Bulletin of the Calcutta Mathematical Society, 56, (1964), 147–162].

35. Conformally Euclidean spaces of class one [Indian Journal of Mathematics, 6, (1964), 93–104].

36. Sir Asutosh Mookerjee—Life sketch, mathematical papers and their summaries [Bulletin of the Calcutta Mathematical Society, 56, (1964), 49–62].

37. On an algebraic system of Riemannian spaces [Journal of the Indian Mathematical Society, 29, (1965), 169–185].

38. Generalization of Clifford's parallelism [Tensor (New Series), 16, (1965), 230–242].

39. (With Bandana Gupta) On orthogonal ennuples in a pair of Riemannian spaces [Proceedings of the National Institute of Sciences, India, 32, (1966), 210–216].

40. On a characterization of conformally flat Riemannian spaces of class one [Journal of the Australian Mathematical Society, 6, (1966), 172–178].

41. On conformally flat Riemannian spaces of class one [Proceedings of the American Mathematical Society, 17, (1966), 880–883].

42. Postulational method in the development of Mathematics [Everyman's Science (Indian Science Congress Association), I, No. I, (1966), 31–34].

43. Application of an algebraic system in Finsler Geometry [Tensor (New Series), 18, (1967), 191–195].

44. On curvature tensors in Finsler Geometry [Tensor (New Series), 18, (1967), 217–226].

45. (With M. C. Chaki) On curvature restrictions of a certain conformally kind of conformally flat Riemannian spaces of class one [Proceedings of the National Institute of Sciences, India, 33, (1967), 100–102].

46. Some generalized formulae for curvature tensor in Finsler Geometry [Indian Journal of Mathematics, 9, (1967), 211–221].

47. Generalized curvature tensors [Bulletin of the Calcutta Mathematical Society, 59, (1967), 9–17].

48. Finsler spaces of recurrent curvature [Tensor (New Series), 19, No. 3, (1968), 291–299].

49. United tensor and connection in the new theories of space [Sir Asutosh Mookerjee Centenary Volume, Indian Association for the Cultivation of Science, Jadavpur, (1968), 35–44].

50. Basic vectors of generalized Riemannian space [Indian Journal of Mechanics and Mathematics, Pt. 1, Sp. Issue, (1968), 95–104].

51. Correction to a theorem of mine [Proceedings of the American Mathematical Society, 27, No. 2, (1971), 341–342].

52. On affine connections in Riemannian, almost complex and almost Hermite spaces [Tensor (New Series), 25, (1972), 390–394].

53. Cyclic structures of geometric objects involving a connection and Lie derivatives [Colloquium Mathematics, 26, (1972), 249–261].

List of Publications of Professor Bibhuti Bhusan Sen

1. Flexure of a beam having a section in the form of a right-angled triangle [Bulletin of the Calcutta Mathematical Society, 21, (1929), 181–186].

2. On stresses in circular rings under the action of isolated forces on the rim [Bulletin of the Calcutta Mathematical Society, 22, (1930), 27–38].

3. Stresses due to a small elliptic hole or a crack on the normal axis of a deep beam under constant bending moment [Philosophical Magazine, 12, (1931), 312–319].

4. On the stresses in an elastic sphere having certain discontinuous distributions of normal pressures on the surface [Bulletin of the Calcutta Mathematical Society, 23, (1930), 67–76].

5. On the uniqueness of solution of problems of elasticity connected with the bending of thin plates under normal pressures [Philosophical Magazine, 16 (7), (1933), 975–979].

6. On the effect of small cavities and cracks in a cylinder twisted by torsional and shearing stresses [Z. fur. Angewandte Math. Und Mech., 13, (1933), 374–379].

7. Uber Dreshscwingungen ton Kegligen Staben [Z. Fur. Tech. Phys.].

8. On concentration of stresses due to a small spherical cavity in a uniform beam bent by terminal couples [Bulletin of the Calcutta Mathematical Society, 25, (1933), 107–114].

9. On bending of certain loaded plates [Indian Phys. Math. Journal, 5, (1934), 17–20].

10. Note on some two-dimensional problems of elasticity connected with plates having triangular boundaries [Bulletin of the Calcutta Mathematical Society, 26, (1934), 65–72].

11. Note on the stresses in some rotating circular disks of varying thickness [Philosophical Magazine, 19 (7), (1935), 1121–1125].

12. Note on the bending of circular disks under certain non-uniform distribution of normal thrust [Philosophical Magazine, 20 (7), (1935), 1158–1163].

13. On torsional vibrations of cylindrical rods under variable forces [Indian Phys. Math. Journal, 6, (1935), 41–44].

14. Note on the application of trilinear coordinates in some problems of elasticity and hydrodynamics [Bulletin of the Calcutta Mathematical Society, 27, (1935), 73–78].

15. Note on the stability of a thin plate under edge thrust, the buckling being resisted by a small force varying as the displacement [Bulletin of the Calcutta Mathematical Society, 27, (1935), 157–164].

16. On the stresses in some solids of revolution due to frictional forces acting on their curved surfaces [Indian Phys. Math. Journal, 7, (1936), 11–15].

17. On the radial vibration of spheres under variable radial forces [Indian Phys. Math. Journal, 7, (1936), 43–46].

18. Note on the transverse vibration of freely supported rectangular plates under the action of moving loads and variable forces [Bulletin of the Calcutta Mathematical Society, 28, (1936), 199–208].

19. Note on the torsion of a curved rod [Philosophical Magazine, 24 (7), (1937), 203–272].

20. Die spainnunge in dunnen halbkreisformrngen aund halbellipetis chen scheiben die un den sie begrenzenden durchemesser rotiere [Z. fur. Angewandte Math. Und Mech., 17, (1937), 181–183].

21. On the transverse vibration of some rotating rods of variable cross section [Indian Phys. Math. Journal, 8, (1937), 49–54].

22. On the stresses produced by couples in layer of elastic material [Bulletin of the Calcutta Mathematical Society, 29, (1937), 41–48].

23. Note on the torsion of a curved rod of circular section [Bulletin of the Calcutta Mathematical Society, 29, (1937), 99–108].

24. Direct determination of stresses from the stress-equations in some two-dimensional problems of elasticity, Part I [Philosophical Magazine, 26 (7), (1938), 98–119].

25. Direct determination of stresses from the stress-equations in some two-dimensional problems of elasticity, Part II, Thermal stresses [Philosophical Magazine, 27 (7), (1939), 437–444].

26. Direct determination of stresses from the stress-equations in some two-dimensional problems of elasticity, Part III, Problems of non-isotropic material [Philosophical Magazine, 27 (7), (1939), 596–604].

27. Stresses due to forces and couples acting in the interior of semi-infinite, elastic solid [Bulletin of the Calcutta Mathematical Society, 32, (1940), 72–83].

28. Note on the bending of thin uniformly loaded plates bounded by cardioids, lemniscates and certain other quartic curves [Philosophical Magazine, 33 (7), (1942), 294–302].

29. Stresses in an infinite strip due to an isolated couple acting at a point inside it [Bulletin of the Calcutta Mathematical Society, 34, (1942), 45–51].

30. Stresses due to forces and couples acting in the interior of an infinite elastic slab placed on rigid foundations [Bulletin of the Calcutta Mathematical Society, 35, (1943), 13–20].

31. Note on the uniqueness of solution of problems of thin plate bent by normal pressures [Bulletin of the Calcutta Mathematical Society, 35, (1943), 135–140].

32. Boundary value problems of circular disks under body forces, Part I [Bulletin of the Calcutta Mathematical Society, 36, (1944), 52–62].

33. Boundary value problems of circular disks under body forces, Part II [Bulletin of the Calcutta Mathematical Society, 36, (1944), 83–86].

34. Direct determination of stresses from the stress-equations in some two-dimensional problems of elasticity, Part IV—'Problems of wedges' [Philosophical Magazine, 36 (7), (1945), 66–72].

35. Stresses in an infinite plate due to isolated forces and couples acting near a circular hole [Philosophical Magazine, 36 (7), (1945), 211–218].

36. Problems of thin plates with circular holes [Bulletin of the Calcutta Mathematical Society, 37, (1945), 37–42].

37. Two-dimensional boundary value problems of elasticity [Proceedings of the Royal Society, London, A, 187, (1946), 87–101].

38. Boundary value problems of a heavy circular disc in a vertical plane [Philosophical Magazine, 37 (7), (1946), 66–72].

39. Note on the stresses in a semi-infinite plate produced by a rigid punch on the strained boundary [Bulletin of the Calcutta Mathematical Society, 38, (1946), 117–120].

40. Direct determination of stresses in thin elastic plates having cavities of different shapes [Bulletin of the Calcutta Mathematical Society, 39, (1947), 113–118].

41. Direct determination of stresses from the stress-equations in some two-dimensional problems of elasticity, Part V—'Problems of curvilinear boundaries' [Philosophical Magazine, 39 (7), (1948), 992–1000].

42. Note on the deformation produced by some symmetrical distribution of variable loads, on the plane boundary of a semi-infinite elastic solid [Bulletin of the Calcutta Mathematical Society, 41, (1949), 77–82].

43. Stresses due to nuclei of thermo-elastic strain in a thin circular plate [Bulletin of the Calcutta Mathematical Society, 42, (1950), 253–255].

44. Note on the stresses produced by nuclei of thermo-elastic strain in a semi-infinite elastic solid [Quarterly of Applied Mathematics, 8, (1951), 365–369].

45. Note on two-dimensional indentation problems of non-isotropic semi-infinite elastic medium [Z. fur. Angewandte Math. Phys., 5, (1954), 83–86].

46. Note on the solution of some problems of semi-infinite elastic solids with transverse isotropy [Indian Journal of Theoretical Physics, 2, (1954), 87–90].

47. Note on a type of distorsionless transmission line with variable parameters [Indian Journal of Theoretical Physics, 4, (1956), 85–87].

48. Note on a direct method of solving problems of elastic plates with circular boundaries [Z. fur. Angewandte Math. Phys., 8, (1957), 307–309].

49. Note on some problems of thin equilateral triangular plate [Indian Journal of Theoretical Physics, 5, (1957), 77–79].

50. Direct method of solving some two-dimensional problems of elasticity [Bulletin of the Calcutta Mathematical Society, Golden Jubilee Commemoration Volume, (1958), 173–178].

51. Note on the uniqueness of solution of problems connected with thin plates bent by normal pressures [Indian Journal of Theoretical Physics, 7, (1959), 41–44].

52. Note on the bending of a thin equilateral plate under tension [Z. fur. Angewandte Math. Und Mech., BAND 40, (1960), 276–277].

53. Some problems of indentation on the straight edge of a semi –infinite non-isotropic plate [Proceedings of the National Institute of Sciences, India, 26 A, (1960), 10–13].

54. Note on the transient response of a linear visco-elastic plate in the form of an equilateral triangle [Indian Journal of Theoretical Physics, 10, (1962), 77–81].

55. Note on the direct determination of steady state thermal stresses in circular disks and spheres [Bulletin of the Calcutta Mathematical Society, 56, (1964), 77–81].

56. Note on the flow of viscous liquid through a channel of equilateral triangular section under exponential pressure gradient [Rev. Roum. Sci. Tech. Ser. Mechanique Appl., 9, (1964), 307–310].

57. Note on the problem of a finite rod of visco-elastic Voigt material [Rev. Roum. Sci. Tech. Ser. Mechanique Appl., 16 (6), (1972), 1237–1241].

List of Publications of Professor Raj Chandra Bose

1. (With S. Mukhopadhyaya) General theorem of cointimacy of symmetries of a hyperbolic triad [Bulletin of the Calcutta Mathematical Society, 17, (1926), 39–54].
2. New methods in Euclidean geometry of four dimensions [Bulletin of the Calcutta Mathematical Society, 17, (1926), 105–140].
3. (With S. Mukhopadhyaya) Triadic equations in hyperbolic geometry [Bulletin of the Calcutta Mathematical Society, 18, (1927), 99–110].
4. The theory of associated figures in hyperbolic geometry [Bulletin of the Calcutta Mathematical Society, 19, (1928), 101–116].
5. Theorems in the synthetic geometry of the circle on the hyperbolic plane [Tohuku Mathematical Journal, (Japan), 34, (1931), 42–50].
6. On a new derivation of the fundamental formulas of hyperbolic geometry [Tohuku Mathematical Journal, (Japan), 34 (1931), 291–294].
7. Synthetic relations between any three elements of a right-angled triangle on the hyperbolic plane [Journal of the Indian Mathematical Society, 19, (1931), 126–129].
8. Generalizations of Roeser's correspondence between certain types of polyhydra in non-Euclidean space [Maths. Physic. Journal, 3, (1932), 44–51].
9. (With W. Blaschke) Quadrilateral 4-webs of curves in a plane [Maths. Physic. Journal, 3, (1932), 99–101].
10. Correspondence between a tetrahedron and a special type of heptahedron in hyperbolic space [Maths. Physic. Journal, 3, (1932), 133–137].
11. On the number of circles of curvature perfectly enclosing or perfectly enclosed by a closed convex oval [Maths. Zeitschrift, (Leipzig), 35, (1932), 16–24].
12. Functional equations satisfied by the fundamental functions of hyperbolic geometry and their application to the geometry of the circle [Maths. Physic. Journal, 4, (1933), 37–41].
13. On the application of hyperspace geometry to the theory of multiple correlation [Sankhya, 1 (1934), 338–342].
14. A note on the convex oval [Bulletin of the Calcutta Mathematical Society, 26, (1935), 55–60].
15. A theorem on the non-Euclidean triangle [Bulletin of the Calcutta Mathematical Society, 26 (1935), 69–72].
16. (With S. N. Roy) Some properties of the convex oval with reference to its perimeter centroid [Bulletin of the Calcutta Mathematical Society, 26, (1935), 79–86].
17. (With S. N. Roy) A note on the area centroid of a closed convex oval [Bulletin of the Calcutta Mathematical Society, 26, (1935), 111–118].
18. (With S. N. Roy) On the four centroids of a closed convex surface [Bulletin of the Calcutta Mathematical Society, 26, (1935), 119–147].
19. (With S. N. Roy) On the evaluation of the probability integral of the D^2 statistic [Science and Culture, (Calcutta), (1935), 436–437].

20. On the exact distribution and moment-coefficients of the D^2 statistic [Sankhya, 2, (1936), 143–154].

21. Theory of skew rectangular pentagons of hyperbolic space. Derivation of the set of associated pentagons [Bulletin of the Calcutta Mathematical Society, 28 (1935), 159–186].

22. A note on the distribution of the differences in mean values of samples drawn from two multivariate normally distributed populations and the definition of the D^2 statistic [Sankhya, 2, (1936), 379–384].

23. Two theorems of the convex oval [Journal of the Indian Mathematical Society (New Series), 2, (1936), 13–15].

24. A theorem on equiangular convex polygons circumscribing a convex curve [Journal of the Indian Mathematical Society (New Series), 2, (1936), 96–98].

25. Analogue of a theorem of Blaschke [Journal of the Indian Mathematical Society, (New Series) 2 (1936), 105–106].

26. (With P. C. Mahalanobis and S. N. Roy) Normalization of variates and the use of rectangular coordinates in the theory of sampling distributions [Sankhya, 3, (1936), 1–40].

27. On a criterion for the existence of a cyclic point [Tohuku Mathematical Journal, (Japan) 43, (1936), 84–88].

28. A note on the osculating circles of a plane curve [Bulletin of the Calcutta Mathematical Society, 29, (1935), 29–32].

29. On the distribution of means of samples drawn from a Bessel functions population [Sankhya, 3 (1938), 262–266].

30. On the application of the properties of Galois fields to the problem of construction of hyper Graeco-Latin squares [Sankhya, 3, (1938), 323–339].

31. (With S. N. Roy) Distribution of the Studentized D^2 statistic [Sankhya, 4, (1938), 19–38].

32. (With K. R. Nair) Partially balanced incomplete block designs [Sankhya, 4, (1939), 337–373].

33. (With K. Kishen) On partially balanced Youden squares [Science and Culture, (Calcutta), 4, (1939), 136–137].

34. On the construction of balanced incomplete block designs [Annals Eugenics, (London), 9, (1939), 358–398].

35. (With S. N. Roy) The use and distribution of the Studentized D^2 statistic when the variances and covariances are based on & samples [Sankhya, 4, (1940), 535–542].

36. (With K. Kishen) On the problem of confounding in the general symmetrical factorial design [Sankhya, 5, (1940), 21–36].

37. (With K. R. Nair) On complete sets of Latin squares [Sankhya, 5, (1941), 361–382].

38. Some new series of balanced incomplete block designs [Bulletin of the Calcutta Mathematical Society, 34, (1942), 17–31].

39. An affine analogue of Singer's theorem [Journal of the Indian Mathematical Society (New Series), 6, (1942), 1–15].

40. A note on two combinatorial problems having applications in the theory of design of experiments [Science and Culture, (Calcutta), 8, (1942), 192–193].

41. A note on the resolvability of balanced incomplete block designs [Sankhya, 6, (1942), 105–110].

42. A note on two series of balanced incomplete block designs [Bulletin of the Calcutta Mathematical Society, 35 (1943), 129–130].

43. The fundamental theorem of linear estimation [Proceedings of the Indian Science Congress, (1944), 4–5].

44. (With S. Chowla and C. R. Rao) On the integral order (mod p) of quadratics $x^2 + ax + b$, with applications to the construction of minimum functions for $GF(P^2)$, and to some number theory result [Bulletin of the Calcutta Mathematical Society, 36, (1944), 153–174].

45. (With S. Chowla) On the construction of affine difference sets [Bulletin of the Calcutta Mathematical Society, 37, (1945), 107–112].

46. Mathematical theory of the Symmetrical factorial design [Sankhya, 8, (1947), 107–166].

47. On a resolvable series of balanced incomplete block designs [Sankhya, 8, (1947), 249–256].

48. Recent work on 'Incomplete Block Design' in India [Biometrics, 3, (1947), 176–178].

49. The design of experiments [Presidential Address, Section of Statistics. Proceedings of the 84th Indian Sci. Congress, (1947)].

50. A note on Fisher's inequality for balanced incomplete block designs [Annals of Mathematical Statistics, 20, (1949), 619–620].

List of Publications of Professor Bhoj Raj Seth

1. Motion of a liquid contained in rotating cylinders of triangular cross-section [Quarterly Journal of Mathematics, 5, (1934), 161–171].

2. On flexure of prisms with cross sections of uniaxial symmetry [Proceedings of the London Mathematical Society, 39, (1934), 502–511].

3. Torsion of beams with T and L-cross sections [Proceedings of the Cambridge Philosophical Society, 30, (1934), 392–403].

4. Torsion of beams whose cross-section is a regular polygon of n sides [Proceedings of the Cambridge Philosophical Society, 30, (1934), 139–149].

5. Finite strain in elastic problems—I [Philosophical Transactions of the Royal Society, 234 A, (1935), 231–264].

6. General solutions of a class of physical problems [Philosophical Magazine, 20, (1935), 632–640].

7. On the solution of a simple flexure problem [Journal of the London Mathematical Society, 10, (1935), 105–107].

8. (With W. M. Shephard) Finite strain in elastic problems—II [Proceedings of the Royal Society, London, 156 A, (1936), 171–182].

9. Flexure of beams of polygonal cross-section [Philosophical Magazine, 22 (7), (1936), 582–592].

10. Flexure of a hollow shaft—I [Proceedings of the Indian Academy of Sciences, 4 A, (1936), 531–541].

11. On flexure of beams of triangular cross-section [Proceedings of the London Mathematical Society, Series 2, 41, (1936), 323–331].

12. Variation of double refraction in celluloid with amount of permanent stretch at constant temperature and at different temperatures [Proceedings of the Physics Society, 48, (1936), 48, 477–486].

13. Vortex motion in rectangular cylinder [Proceedings of the Indian Academy of Sciences, 4 A, (1936), 435–441].

14. Flexure of a hollow shaft—II [Proceedings of the Indian Academy of Sciences, 5 A, (1937), 23–31].

15. On the sufficiency of the consistency equations [Proceedings of the Indian Academy of Sciences, 5 A, (1937), 518–521].

16. Symmetrical flexure of an angle-iron [Philosophical Magazine Supplement, (1937), 745–757].

17. On waves in canals of variable depth [Philosophical Magazine Supplement, 23 (7), (1937), 106–114].

18. Waves in a circular channel [Philosophical Magazine Supplement, 24, (1937), 288–293].

19. Transverse waves in canals [Proceedings of the Indian Academy of Sciences, 7 A, (1938), 104–107].

20. Two-dimensional potential problems connected with rectilinear boundaries [Lucknow University Studies, No. 13, (1939), Allahabad Law Journal Press].

21. Application of theory of finite strain [Proceedings of the Indian Academy of Sciences, 9 A, (1939), 17–19].

22. On the motion of a liquid set up by a moving regular polygonal cylinder [Journal of the London Mathematical Society, 14, (1939), 255–261].

23. Potential problems concerning curved boundaries [Proceedings of the Indian Academy of Sciences, 9 A, (1939), 447–453].

24. Potential solutions near an angular point [Proceedings of the Indian Academy of Sciences, 9 A, (1939), 136–138].

25. Some problems of finite strain—I [Philosophical Magazine, 27, (1939), 286–293].

26. Some problems of finite strain—II [Philosophical Magazine, 27, (1939), 449–452].

27. Uniform motion of a sphere or a cylinder through a viscous liquid [Philosophical Magazine, 27, (1939), 212–220].

28. Transverse vibrations of triangular membranes [Proceedings of the Indian Academy of Sciences, 12, (1940), 487–490].

29. On the gravest mode of some vibrating systems [Proceedings of the Indian Academy of Sciences, 13 A, (1941), 390–394].

30. Finite strain in a rotating shaft [Proceedings of the Indian Academy of Sciences, 14 A, (1942), 648–651].

31. On Guest's law of elastic failure [Proceedings of the Indian Academy of Sciences, 14 A, (1942), 37–40].

32. Viscous solutions obtained by superposition of effects [Proceedings of the Indian Academy of Sciences, 16 A, (1942), 193–195].

33. Consistency equation of finite strain [Proceedings of the Indian Academy of Sciences, 20 A, (1944), 336–339].

34. On the strain stress-strain velocity relations in equation of viscous flow [Proceedings of the Indian Academy of Sciences, 20 A, (1944), 329–335].

35. Bending of an equilateral plate [Proceedings of the Indian Academy of Sciences, 22 A, (1945), 234–238.

36. Finite strain in aeolotropic elastic bodies—I [Bulletin of the Calcutta Mathematical Society, 37, (1945), 62–68].

37. Finite strain in aeolotropic elastic bodies—I [Bulletin of the Calcutta Mathematical Society, 38, (1946), 39–44].

38. On Young's modulus for Indian rubber [Bulletin of the Calcutta Mathematical Society, 38, (1946), 143–144].

39. Stability of rectilinear plates [Journal of the Indian Mathematical Society (NS), 10, (1946), 13–16].

40. Bending of clamped rectilinear plates [Philosophical Magazine, 38, (1947), 292–297].

41. Finite longitudinal vibrations [Proceedings of the Indian Academy of Sciences, 25 A, (1947), 151–152].

42. Transverse vibrations of rectilinear plates [Proceedings of the Indian Academy of Sciences, 25 A, (1947), 25–29].

43. Bending of rectilinear plates [Bulletin of the Calcutta Mathematical Society, 40, (1948), 36–40].

44. Bending of an elliptic plate with confocal hole [Quarterly Journal of Applied Mechanics and Mathematics, 2, (1949), 177–181].

45. Some recent applications of the theory of finite elastic deformation [Proceedings of the Symposium on Applied Mathematics, 3, (1950), 67–84, McGraw Hill, New York].

46. Some solutions of the wave equation [Proceedings of the Indian Academy of Sciences, 32 A, (1950), 421–423].

47. Synthetic method for non-linear problems [Proceedings of the International Congress on Mathematics, 5 (1), (1950), 636–637].

48. Boundary conditions interpreted as conformal transformation [Proceedings of the American Mathematical Society, 2, (1951), 1–4].

49. Finite elastic-plastic torsion [Journal of Mathematics and Physics, 31, (1952), 84–90].

50. Generalized singular points with applications to flow problems [Proceedings of the Indian Academy of Sciences, 40 A, (1954), 25–26].

51. Hydrodynamical generalized singular points [Current Science, 23, (1954), 148–185].

52. New formulation of equations of incompressible flow [Bulletin of the Calcutta Mathematical Society, 46, (1954), 217–220].

53. Synthetic method for incompressible flow [Proceedings of the International Congress on Mathematics, Amsterdam, (1954), 4].

54. Wave motion and vibration theory [Current Science, 23, (1954), 389–390; Proceedings of the 5th Symposium of the American Mathematical Society, McGraw Hill, New York, (1954), 169].

55. Non-linear Continuum Mechanics [Presidential address Mathematics Section, Indian Science Congress Association, Baroda, (1955), 21–48].

56. Stability of rectilinear plates [Z. Angew. Math. Mech., 35, (1955), 96–99].

57. Elastic and fluid flow problems for triangular and quadrilateral boundaries [Proceedings of the International Congress on Theoretical and Applied Mechanics, 5, Brussels, (1956)].

58. Finite bending of a plate into a spherical shell [Z. Angew. Math. Mech., 37, (1957), 393–398].

59. Finite thermal strain in spheres and circular cylinders [Arch. Mech. Stos, 9, (1957), 633–645].

60. New solutions for finite deformation [Proceedings of the Indian Academy of Sciences, 45 A, (1957), 105–112].

61. Synthetic method for boundary layer thickness [Proceedings of the IUTAM Symposium on Boundary Layer Research, Freiburg, (1957), 47–48].

62. Finite bending of a non-homogeneous aeolotropic sheet (elastic bending) [Proceedings of the Indian Congress on Theoretical and Applied Mechanics, 4, (1958), 38–44].

63. Finite strain in engineering design [Les mathematiques de lingenier, (1958), 386–390, Men Publ. Soc. Sci., Arts Lett., Hainant, Vol. I].

64. Non homogeneous yield conditions: Non-homogeneity in elasticity and plasticity [Proceedings of the IUTAM Symposium, Warsaw; Pergamon Press, London (1958)].

65. Non-linear rotational flows [Proceedings of the Indian Congress on Theoretical and Applied Mechanics, 3, (1958), 199–202].

66. Pro-plastic deformation [Sonderdruck aus Rhelogica Acta, 2/3, (1958), 316–318].

67. Paraboloidal bending of plates [Proceedings of the Indian Congress on Theoretical and Applied Mechanics, 5, (1959), 99–102].

68. Boundary layer research [Journal of the Indian Mathematical Society, 24, (1960), 527–550].

69. Boundary layer thickness [Journal of Science and Engineering Research, 4, (1960), 1–6].

70. Finite bending of plates into cylindrical shells [Annals of Mathematics Pure Appl., 50, (1960), 119–125].

71. Finite deformation of cylindrical shells [Proceedings of the International Symposium on Thin Elastic Sheets, Delft, (1960), 355–362].

72. Finite deformation of plates into shells [In: Partial Differential equations and continuum mechanics, (1961), 95–105. University of Wisconsin Press, USA].

73. Problems of continuum mechanics: Contributions in honour of the Seventieth Birthday of Academician N. I. Muschkelisvili, Society for Industrial and Applied Mathematics, Philadelphia, (1961), 60].

74. Stability of finite deformation problems of continuum mechanics [Muschkelisvili Anniversary Volume, (1961), 406–413, SIAM, Philadelphia].

75. Elastic plastic transition in shells and tubes under pressure [Z. Angew. Math. Mech., 43, (1962), 345–351].

76. Generalized strain measure with application to physical problems: second order effects in elasticity, plasticity and fluid dynamics [Proceedings of the IUTAM Symposium, Haifa (1962), 162–172, Academic Press].

77. On a functional equation in finite deformation [Z. Angew. Math. Mech., 42, (1962), 391–396].

78. Simple case of transition phenomenon [Proceedings of the Army Mathematical Conference, 8, (1962), Madison, 409–447].

79. Transition theory of elastic plastic deformation, creep and relaxation [Nature, 195, (1962), 896–897].

80. Asymptotic phenomena in large rotation [Journal of Mathematics and Mechanics, Indiana, 12, (1963), 205–212].

81. Asymptotic phenomena in large torsion [Journal of Mathematics and Mechanics, Indiana, 12, (1963), 193–204].

82. Fifty years of Science in India: Progress of Mathematics [Indian Science Congress Association, Calcutta, (1963)].

83. International symposium on continuum mechanics [Journal of Scientific and Industrial Research, 23, (1963), 1–7].

84. Mixed boundary value problems [Calcutta Mathematical Society Golden Jubilee Commemoration Volume, (1963), 79–86].

85. Transition theory of strip bending [Journal of Applied Mathematics and Mechanics, 27, (1963), 571–576].

86. Transition theory of elastic-plastic deformation [A course of extension lectures delivered at Osmania University, Hyderabad, India, (1964), Bangalore Press, Bangalore].

87. Elastic-plastic transition in torsion [Z. Angew. Math. Mech., 44, (1964), 229–233].

88. Asymptotic treatment of transition problems in mechanics [Proceedings of the Indian Congress on Theoretical and Applied Mechanics, 9, (1964), 1–3].

89. Eleventh International Congress on Applied Mechanics [Journal of Scientific and Industrial Research, 23, (1964), 499–501].

90. On the problem of transition phenomena—I [Bulletin of Institute Polytechnic Din IASI, 10, (1964), 255–262].

91. Survey on second-order elasticity. Second-order effects in elasticity, plasticity and fluid dynamics [Pergamon Press, Oxford, (1964), 261].

92. Transition phenomena in physical problems [Sir Asutosh Mookerjee Birthday Commemoration Volume of the Bulletin of the Calcutta Mathematical Society, 56, (1964), 83–89].

93. Application of the theory of functions in continuum mechanics [Proceedings of the International Symposium, Toitisi, Moscow, 2, (1965), 382–388].

94. Continuum concepts of measure [Presidential address. Proceedings of the Indian Congress on Theoretical and Applied Mechanics, 10, (1965), 1–15].

95. Generalized singular points with applications to flow problems [Proceedings of the Indian Academy of Sciences, 40 A, (1965), 25–36].

96. Subharmonic problems of continuum mechanics: applied theory of function in continuum mechanics—I, Proceedings of the International Symposium, Toitisi, (1963), 2. Fluid and Gas Mechanics, Mathematical Methods (Russian), (1965), Izdat, Nauka, Moscow.

97. Bendiñg of T Plate, Proceedings of the Indian Congress on Theoretical and Applied Mechanics, 11, (1966), 87–90.

98. Measure concept in mechanics [International Journal of Nonlinear Mechanics, 1, (1966)].

99. Plane transitions [Indian Journal of Mathematics, 9, (1967), 499–504].

100. Irreversible transition in continuum mechanics: Irreversible aspects in continuum mechanics, Physical characteristics in moving fluids [Proceedings of the IUTAM Symposium, Vienna, (1968), 359–366].

101. On the deformation of elastic viscoplastic bodies: Problems of Hydrodynamics and Continuum Mechanics [Contribution in honour of Sixtieth Birthday of Academician L. I. Sedov, SIAM, (1969), Philadelphia].

102. Transition conditions: the yield condition [International Journal of Nonlinear Mechanics, (GB), 5, (1970), 279–285].

103. Transition on analysis of collapse of thick cylinders [Z. Angew. Math. Mech., 50, (1970), 617–621].

104. Creep transition [Journal of Mathematical and Physical Sciences (India), 6, (1972), 73–81].

105. Creep transition in continuum mechanics and related problems. [Academician N. I. Muschkelisvili Eightieth Birthday Commemoration Volume, Izdat, Nauka, Moscow, (1972), 459–464].

106. Yield conditions in plasticity [Arch. Math. Stosow, Poland, 24, (1973), 769–776].

107. Creep plastic effects in sheet bending [Z. Angew. Math. Mech., 54, (1974), 557–561].

108. Creep transition in rotating cylinders [Journal of Mathematical and Physical Sciences (India), 8, (1974), 1–5].

List of Publications of Professor Subodh Kumar Chakrabarty

1. Stark-Effekt des rotationsspektrums and electrischesuzeptibilitatbeihoher temperature [Verlag Von Julius Springer, Zeitschrift fur Physic, Berlin Sonderbdruck, 102, Band 1 and 2 Heft (1936), 102–111].

2. Das eigenwert-problem, eineszweiartomigenmolekuls und die berechung der dissoziationsenergie [Verlag Von Julius Springer, Zeitschrift fur Physic, Berlin Sonderbdruck, 109, Band 1 and 2 Heft (1937), 25–38].

3. Notizuber den stark effect der rotationsspektren [Verlag Von Julius Springer, Zeitschrift fur Physic, Berlin Sonderbdruck, 110, Band 11, 12 Heft (1938), 688–691].

4. Quantization under two centres of forces—part I the hydrogen molecular ion [Philosophical Magazine, 7, XXVIII (1939), 423–434].

5. Production of bwests by meson and its dependence on the meson spin [Indian Journal of Physics, XVI (VI), (1942), 377–392].

6. (With H. J. Bhabha) The cascade theory with collision loss [Proceedings of the Indian Academy of Science, A, 15, (1942), 462].

7. (With H. J. Bhabha) The cascade theory with collision loss [Proceedings of the Indian Academy of Science, A, 15, (1942), 464–476].

8. Accurate calculations on the cascade theory of electronic showers without collision loss [Proceedings of the National Institute of Science (India), 8, (1942), 331].

9. (With H. J. Bhabha) The cascade theory with collision loss [Proceedings of the Royal Society, London, A, 181, (1943), 267–303].

10. The atmospheric absorption curves and their dependence on the nature of primary cosmic rays [Indian Journal of Physics, XVII (VI), (1943), 121–129].

11. The effect of screening on the bremsstrahlung and pair-creation process and its consequence on the cascade theory [Proceedings of the National Institute of Science (India), 9 (2), (1943), 323–335].

12. On the convergency of the solutions of cascade equations in cosmic radiation [Bulletin of the Calcutta Mathematical Society, 36, (1944), 9–13].

13. Photons associated with a cascade shower [Bulletin of the Calcutta Mathematical Society, 36, (1944), 135–140].

14. (With R. C. Majumdar) On the spin of the meson [Physical Review, 65, (1944), 206].

15. Shower production by mesons in cosmic radiation [Bulletin of the Calcutta Mathematical Society, 37(3), (1945), 95–106].

16. Solar stream of corpuscles and their relation to magnetic storms [Monthly Notices of the Royal Astronomical Society, 106, (1946), 491–499].

17. Cascade showers under thin layers of materials [Nature, 158, (1946), 166].

18. Geomagnetic time variations and their relation to ionospheric conditions [Current Science, 15, (1946), 246–247].

19. Frequency of micropulsations and their variations at Alibag [Government of India Publications (Simla), X (126), (1946), 147–152].

20. Generation of mesons and its dependence on meson spin [Bulletin of the Calcutta Mathematical Society, 39(4), (1947), 166–176].

21. Scientific Notes [Bulletin of the Indian Meteorological Department, 10, (1947), 126].

22. (With H. J. Bhabha) Further calculation on the cascade theory [Physical Review, 74 (10), (1948), 1352–1363].

23. (With C. F. Richter) The Walker Pass earthquakes and structure of the Southern Sierra Nevada [Bulletin of the Seismological Society of America, 39 (2), (1949), 93–107].

24. Response characteristics of electromagnetic seismographs and their dependence on the instrumental constants [Bulletin of the Seismological Society of America, 39 (3), (1949), 205–2018].

25. Sudden commencements in geomagnetic field variations [Nature, 167, (1951), 31].

26. (With R. Pratap) On the dynamo theory of the geomagnetic field variations [Journal Geophysical Research, 59 (1), (1954), 1–14].

27. The spherical harmonic analysis of the Earth's main magnetic field [Indian Journal of Meteorology and Geophysics, 5, (1954), 63–68].

28. (With M. R. Gupta) Calculations of the cascade theory of showers [Physical Review, 101 (2), (1956), 813–819].

29. Cosmic Ray works at the B. E. College, Howrah [Journal of the Scientific and Industrial Research, 17 A (12), Supplement, (1958), 81–82].

30. Contribution of electron tridents in cascade showers [Bulletin of the Calcutta Mathematical Society, The Golden Jubilee Commemoration Volume, (1958), 217–223].

31. (With D. Sarkar) Microseisms associated with Nor'westers [Bulletin of the Seismological Society of America, 48, (1958), 181–189].

32. Response characteristics of electromagnetic seismographs [Proceedings of the National Institute of Science (India), 26 A, (Suppl. II), (1960), 133–142].

33. Response characteristics of electromagnetic seismographs [Proceedings of the National Institute of Science (India), Silver Jubilee Volume (1960)].

34. (With A. N. Tandon) Calibration of electromagnetic seismographs satisfying Galitz conditions [Bulletin of the Seismological Society of America, 51 (1), (1961), 111–125].

35. (With S. N. Roychoudhury) Response characteristics of electromagnetic seismographs [Bulletin of the Seismological Society of America, 54 (5), (Part A) (1964), 1445–1458].

36. (With G. C. Choudhury and S. N. Roychoudhury) Magnification curves of electromagnetic seismographs [Bulletin of the Seismological Society of America, 54 (5), (Part A) (1964), 1459–1471].

37. Propagation of waves in a multilayered elastic medium and their dependence on source mechanism [Bulletin of the Seismological Society of America, 57 (6), (1967), 1449–1465].

38. On hypocentral dislocations and associated seismic waves [Proceedings of the Symposium of Geophysics, Calcutta, (1969), 47–54].

39. Oscillations of the Earth and their use in the study of the internal constitution of the Earth [Bulletin of the Calcutta Mathematical Society, 63, (1971), 11–17].

List of Publications of Professor Manindra Chandra Chaki

1. Some formulas in Tensor calculus, [Bulletin of the Calcutta Mathematical Society, 42, (1950), 249–252].
2. On a non-symmetric harmonic space, [Bulletin of the Calcutta Mathematical Society, 44, (1952), 37–40].
3. (With H. Bagchi): Note on certain remarkable types of plane collineations, [Ann Scuola Norm Super, Pisa (3), 6, (1952), 85–97].
4. (With H. Bagchi): Note on autopolar plane cubics, [Rendi Sem Math Univ. di Padova, 21, (1952), 316–334].
5. (With H. Bagchi): Note on collineation-group connected with a Plane Quadrangle, [Proceedings of the National Academy of Sciences (India), 23(3), (1954)].
6. On the line geometry of a curvature tensor, [Bulletin of the Calcutta Mathematical Society, 47. (1955), 217–226].
7. Some formulas in a Riemannian space, [Ann Scuola Norm Super, Pisa (3), 10, (1956), 85–90].
8. On a type of tensor in a Riemannian space, [Proceedings of the National Institute of Sciences (India), Pt. A, 22, (1956), 89–97].
9. (With B. Gupta): On conformally symmetric spaces, [Indian Journal of Mathematics, 5(2), (1963), 113–122].
10. (With R. N. Sen): On curvature restrictions of a certain kind of conformally-flat Riemannian space of class one, [Proceedings of the National Institute of Sciences (India), Pt. A, 33, (1967), 100–102].
11. (With A. N. Roy Chowdhury): On Ricci-recurrent space of second order, [Indian Journal of Mathematics, 9, (1968), 279–287].
12. (With A. N. Roy Chowdhury): On conformally recurrent spaces of second order, [Journal of the American Mathematical Society, 10, (1969), 155–461].
13. (With D. Ghosh): On a type of Sasakian space, [Journal of the American Mathematical Society, 13, (1972), 508–510].
14. On conformally recurrent Kahler spaces, [Commemoration volumes for Prof. Dr. Akitsugu Kawaguchi's Seventieth birthday, Vol. Il, Tensor (New Series), 25, (1972), 179–182].
15. (With D. Ghosh): On a type of K-contact Riemannian manifold, [Journal of the American Mathematical Society, 13, (1972), 447–450].
16. (With A. N. Roy Chowdhury): On a type of Kahler space, [Mathematics(Cluj), 16(39), (1974), No. 2, 223–227].
17. (With A. K. Ray): On conformally flat generalized 2-recurrent spaces, [Publication Mathematics Debrecen, 22, (1975). No. 1–2, 95–99].
18. In Memoriam: Rabindra Nath Sen (1896–1974), [Bulletin of the Calcutta Mathematical Society, 67, (1975), No. 4, 251–257].
19. (With K. K. Sharma): A particular conformally symmetric space, [Colloquium Mathematicum, 35, (1976), No. 1, 87–90].

20. (With K. K. Sharma): Corrections to 'A particular conformally symmetric space', [Colloquium Mathematicum, 38(1), (1977), 169].

21. (With D. Ghosh): On conformally 2-recurrent spaces, [Mat Vesnik, 1(14), (29), No. 1, (1977), 21–23].

22. (With A. K. Ray): On certain types of Kahler spaces, [Publication Mathematics Debrecen, 26(3–4), (1979), 255–262].

23. (With U. C. De): On a type of Riemannian manifold with conservative conformal curvature tensor, [Ann. Soc. Sc. Bruxelles, 195(2), (1981), 81–84].

24. (With U. C. De): On a type of Riemannian manifold with conservative conformal curvature tensor, [Compte Rend. Acad. Bulgari des Sc., 34(7), (1981), 965–968].

25. (With A. Konar): On a type of semisymmetric connection on a Riemannian manifold, [Journal of Pure Mathematics, 1, (1981), 77–80].

26. (With S. K. Kar): On a type of semisymmetric metric connection on a Riemannian manifold, [Compte Rend. Acad. Bulgari des Sc., 36(1), (1983), 57–60].

27. (With M. Tarafdar): On a type of sasakian manifold, [Bull. Moth. Soc. Sci. Math. R, S. Roumanie, (New Series), 27(75), No. 3, (1983), 217–220].

28. (With G. Kumar): On a type of semisymmetric connection on a Riemannian manifold, [An. Stiint. Univ. AI. Cuza Isai Sect I a Mat, 29(2), Suppl., (1983), 41–44].

29. (With G Kumar): On semi-decomposable generalized projective 2-recurrent Riemannian spaces, [Mathematics(Cluj), 26(49), No. 1, (1984), 21–28].

30. (With S. K. Kar): On a type of semisymmetric metric connection on a Riemannian manifold, [Journal of Pure Mathematics, 4, (1984), 102–107].

31. (With B. Chaki): On a type of conformally flat nearly Kahler manifold, [An. Stiint. Univ. AI. Cuza Isai Sect I a Mat, 31(3), (1985), 235–238].

32. (With B. Chaki): On pseudo symmetric manifold admitting a type of semisymmetric connection, [Soochow Journal of Mathematics, 13(1), (1987), 1–7].

33. On pseudo symmetric manifolds, [An. Stiint. Univ. AI. Cuza Isai Sect I a Mat, 33(1), (1987), 53–58].

34. (With M. Tarafdar): On conformally flat pseudo-Ricci symmetric manifolds, [Periodica Mathematica Hungarica 19(3) (1988), 209–215].

35. (With G Kumar): On semi-decomposable generalized conformally 2-Recurrent Riemannian spaces, [Mathematics(Cluj), 30(53), No. 1, (1988), 11–18].

36. On pseudo Ricci symmetric manifolds, [Bulgarian Journal of Physics, 15(6), (1988), 526–531].

37. (With U. C. De): On pseudo symmetric spaces, [Acta Mathematica Hungarica 54(3–4), (1989), 185–190].

38. (With S. K. Saha): On pseudo projective symmetric manifolds, [Bull. Inst. Math. Acad. Sinica, 17(1), (1989), 59–65].

39. (With M. Tarafdar): On a type of sasakian manifold, [Soochow Journal of Mathematics, 16(1), (1990), 23–28].

40. Syamadas Mukhopadhyay (1866–1937), [Journal of Pure Mathematics, 7, (1990), 59–65].

41. (With P. Chakrabarti): On conformally flat pseudo-Ricci symmetric manifolds, [Tensor (New Series), 52(3), 1993], 217–222].

42. (With G Kumar): On semi-decomposable generalized projective 2 Recurrent Riemannian spaces, [Univ. Nac. Tucman Rev., Ser. A 30(1–2), (1993), 129–139].

43. (With S. Koley): On generalized pseudo Ricci symmetric manifolds, [Periodica Mathematica Hungarica, 28(2), (1994), 123–129].

44. On generalized pseudosymmetric manifolds [Publication Mathematics Debrecen, 45(3–4), (1994), 305–312].

45. (With S. K. Saha): On pseudo-projective Ricci symmetric manifolds, [Bulgarian Journal of Physics, 21(1–2), (1994–95), 1–7].

46. (With S. Koley): On generalized pseudo-projective Ricci symmetric manifolds, [An. Stiint. Univ. AI. Cuza Isai Sect I a Mat, 41(1), (1995–96), 75–84].

47. (With S. Ray): Space-times with covariant-constant energy-momentum tensor, [Indian Journal of Theoretical Physics, 35(5), (1996), 1027–1032].

48. (With S. P. Mondal): On generalized pseudosymmetric manifolds, [Publication Mathematics Debrecen, 51(1–2), (1997), 35–42].

49. (With M. L. Ghosh): On quasi-conformally flat and quasi-conformally conservative Riemannian manifolds, [An. Stiint. Univ. AI. Cuza Isai Sect I a Mat (NS), 43(2), (1997), 375–381].

50. (With R. K. Maity): On totally umbilical hypersurfaces of a conformally flat pseudo Ricci symmetric manifold, [Tensor (New Series), 60(3), (1998), 254–257].

51. (With P. Chakrabarti): On a type of conformally flat hyper surface of a Euclidean manifold, [Tensor (New Series), 61(1), (1999), 7–13].

52. On statistical manifolds, [Tensor (New Series), 61(1), (1999), 14–17].

53. (With R. K. Maity): On quasi Einstein manifolds, [Publication Mathematics Debrecen, 57(3–4), (2000), 297–306].

54. (With B. Barua): Symmetric of the synge metric in the relativistic optics, [Bulletin of the Calcutta Mathematical Society, 92(3). (2000), 219–224].

55. (With M. L. Ghosh): On quasi Einstein Manifolds. [B. N. Prasad birth centenary commemoration Volume 11, [Indian Journal of Mathematics, 42(2), (2000), 211–220].

List of Publications of Professor Calyampudi Radhakrishna Rao

1. (With K. R. Nair) Confounded designs for asymmetrical factorial experiments [Science and Culture, 6, (1941), 313–314].

2. On the volume of a prismoid in n-space and some problems in continuous probability [The Mathematics Student, 10, (1942), 68–74].

3. (With K. R. Nair) Confounded designs for $k \times pm \times qn \times \ldots$ type of factorial experiments [Science and Culture, 7, (1942), 361].

4. (With K. R. Nair) A general class of quasi-factorial designs leading to confounded designs for factorial experiments [Science and Culture, 7, (1942), 457–458].

5. (With K. R. Nair) A note on partially balanced incomplete block designs [Science and Culture, 7, (1942), 568–569].

6. On the sum of n observations from different gamma type populations [Science and Culture, 7, (1942), 614–615].

7. (With K. R. Nair) Incomplete block designs for experiments involving several groups of varieties [Science and Culture, 7, (1942), 615–616].

8. On bivariate correlation surfaces [Science and Culture, 8, (1942), 236–237].

9. Certain experimental arrangements in quasi-Latin squares [Current Science, 12, (1943), 322].

10. (With R. C. Bose and S. Chowla) On the integral order (mod p) of quadratics $x^2 + ax + b$, with applications to the construction of minimum functions for $GF(p^2)$ and to some number theory results [Bulletin of the Calcutta Mathematical Society, 36, (1944), 153–174].

11. On linear estimation and testing of hypothesis [Current Science, 13, (1944), 154–155].

12. On balancing parameters [Science and Culture, 9, (1944), 554–555].

13. Extension of the difference theorems of Singer and Bose [Science and Culture, 10, (1944), 57].

14. On the linear set up leading to intra and inter block information [Science and Culture, 10, (1944), 259–260].

15. Information and the accuracy attainable in the estimation of statistical parameters [Bulletin of the Calcutta Mathematical Society, 37, (1945), 81–91]. Familiar correlations or the multivariate generalisation of the interclass correlation [Current Science, 14, (1945), 66–67] (With R. C. Bose and S. Chowla) A chain of congruences [Proceedings of the Lahore Philosophical Society, 7 (1), (1945), 53].

16. (With R. C. Bose and S. Chowla) Minimum functions in Galois fields [Proceedings of the National Academy of Sciences, (India), A 15, (1945), 193].

17. Finite geometries and certain derived results in theory of numbers [Proceedings of the National Institute of Sciences, (India), 11, (1945), 136–149] Generalisation of Markoff's theorem and tests of linear hypotheses [Sankhyā, 7, (1945), 16–19].

18. Studentised tests of linear hypotheses [Science and Culture, 11, (1945), 202–203].

19. Hypercubes of strength 'd' leading to confounded designs in factorial experiments [Bulletin of the Calcutta Mathematical Society, 38, (1946), 67–68].

20. On the mean conserving property [Proceedings of the Indian Academy of Sciences, Sect. A, 23, (1946), 165–173].

21. Difference sets and combinatorial arrangements derivable from finite geometries [Proceedings of the National Institute of Sciences, (India), 12, (1946), 123–135].

22. On the linear combination of observations and the general theory of least squares [Sankhyā, 7, (1946), 237–256].

23. Confounded factorial designs in quasi–Latin squares [Sankhyā, 7, (1946), 295–304].

24. Tests with discriminant functions in multivariate analysis [Sankhyā, 7, (1946), 407–414].

25. (With S. Janardhan Poti) On locally most powerful tests when alternatives are one-sided [Sankhyā, 7, (1946), 439].

26. On the most efficient designs [Sankhyā, 7, (1946), 440].

27. Note on a problem of Ragnar Frisch [Econometrica, 15, (1947), 245–249].

28. General methods of analysis for incomplete block designs [Journal of the American Statistical Association, 42, (1947), 541–561].

29. The problem of classification and distance between two populations [Nature, 159, (1947), 30].

30. A statistical criterion to determine the group to which an individual belongs [Nature, 160, (1947), 835–836].

31. Minimum variance and the estimation of several parameters [Proceedings of the Cambridge Philosophical Society, 43, (1947), 280–283].

32. Factorial experiments derivable from combinatorial arrangements of arrays [Suppl. Journal of the Royal Statistical Society, 9, (1947), 128–139].

33. (With D. C. Shaw) On a formula for the prediction of cranial capacity [Biometrics, 4, (1948), 247–253].

34. Tests of significance in multivariate analysis [Biometrika, 35, (1948), 58–79].

35. (With K. R. Nair) Confounding in asymmetrical factorial experiments [Journal of the Royal Statistical Society, Ser. B, 10, (1948), 109–131].

36. The utilization of multiple measurements in problem of biological classification (with discussion) [Journal of the Royal Statistical Society, Ser. B, 10, (1948), 159–203].

37. Large sample tests of statistical hypotheses concerning several parameters with applications to problems of estimation [Proceedings of the Cambridge Philosophical Society, 44, (1948), 50–57].

38. (With Patrick Slater) Multivariate analysis applied to differences between neurotic groups [British Journal of Psychological Statistics, Sect., 2, (1949), 17–29].

39. Sufficient statistics and minimum variance estimates [Proceedings of the Cambridge Philosophical Society, 45, (1949), 213–218].

40. On a class of arrangements [Proceedings of the Edinburgh Mathematical Society, (2) 8, (1949), 119–125].

41. (With P. C. Mahalanobis) Statistical analysis (Part II of 'Anthropometric Survey of United Provinces', 1941) [Sankhyā, 9, (1949), 111–180].

42. A note on the use of indices (Appendix 2 of 'Anthropometric Survey of United Provinces', 1941) [Sankhyā, 9, (1949), 246–248].

43. On the distance between two populations (Appendix 3 of 'Anthropometric Survey of United Provinces', 1941) [Sankhyā, 9, (1949), 246–248].
44. Representation of 'p' dimensional data in lower dimensions (Appendix 4 of 'Anthropometric Survey of United Provinces', 1941) [Sankhyā, 9, (1949), 248–251].
45. On a transformation useful in multivariate computations (Appendix 5 of 'Anthropometric Survey of United Provinces', 1941) [Sankhyā, 9, (1949), 251–253].
46. On some problems arising out of discrimination with multiple characters [Sankhyā, 9, (1949), 336–360].
47. A note on unbiased and minimum variance estimates [Calcutta Statistical Association Bulletin, 3, (1950), 36].
48. Methods of scoring linkage data giving the simultaneous segregation of three factors [Heredity, 4, (1950), 37–59].
49. The theory of fractional replication in factorial experiments [Sankhyā, 10, (1950), 81–86].
50. Statistical inference applied to classificatory problems, Part I, Null hypothesis, discriminatory problems and distance power tests [Sankhyā, 10, (1950), 229–256].
51. A note on the distribution of $D^2_{p+q} - D^2_p$ and some computational aspects of D^2 statistic and discriminant function [Sankhyā, 10, (1950), 257–268].
52. Sequential tests of null hypotheses [Sankhyā, 10, (1950), 361–370].
53. A simplified approach to factorial experiments and the punched card technique in the construction and analysis of designs [Bulletin Institute International Statist., 33 (2), (1951), 1–28].
54. An asymptotic expansion of the distribution of Wilks' A criterion [Bulletin Institute International Statist., 33 (2), (1951), 177–180].
55. A theorem of least squares [Sankhyā, 11, (1951), 9–12].
56. Statistical inference applied to classificatory problems, Problem II: The problem of selecting individuals for various duties in a specified ration [Sankhyā, 11, (1951), 107–116].
57. The applicability of large sample tests for moving average and autogressive schemes to series of short length—an experimental study, Part 3: The discriminant function approach in the classification of time series (Part III of Statistical inference applied to classificatory problems) [Sankhyā, 11, (1951), 257–272].
58. (With K. Kishen) An examination of various inequality relations among parameters of the balanced incomplete block design [Journal of Indian Society of Agricultural Statistics, 4, (1952), 137–144].
59. Minimum variance estimation in distributions admitting ancillary statistics [Sankhyā, 12, (1952), 53–56].
60. Some theorems on minimum variance estimation [Sankhyā, 12, (1952), 27–42].
61. On statistics with uniformly minimum variance [Science and Culture, 17, (1952), 483–484].

62. Progress of statistics in India: 1939–1950, in 'Progress of Science in India', Section 1: Mathematics, Including Geodesy and Statistics [(Nikhil Ranjan Sen, Ed.), National Institute of Sciences of India, New Delhi, (1953), 68–94].

63. Discriminant functions for genetic differentiation and selection (Part IV of Statistical inference applied to classificatory problems) [Sankhyā, 12, (1953), 229–246].

64. On transformations useful in the distribution problem of least squares [Sankhyā, 12, (1953), 339–346].

65. A general theory of discrimination when the information about alternative population distribution is based on samples [Annals of Math. Statist., 25, (1954), 651–670].

66. Estimation of relative potency from multiple response data [Biometrics, 10, (1954), 208–220].

67. On the use and interpretation of distance functions in statistics [Bulletin Institute International Statist., 34 (2), (1954), 90–97].

68. Estimation and tests of significance in factor analysis [Psychometrica, 20, (1955), 93–111].

69. Analysis of dispersion for multiply classified data with unequal numbers in cells [Sankhyā, 15, (1955), 253–280].

70. (With G. Kallianpur) On Fisher's lower bound to asymptotic variance of a consistent estimate [Sankhyā, 15, (1955), 331–342].

71. (With I. M. Chakravarti) Some small sample tests of significance for a Poisson distribution [Biometrics, 12, (1956), 264–282].

72. Analysis of dispersion with incomplete observations on one of the characters [Journal of the Royal Statistical Society, Ser B 18, (1956), 259–264].

73. On the recovery of interblock information in varietal trials [Sankhyā, 17, (1956), 105–114].

74. A general class of quasifactorial and related designs [Sankhyā, 17, (1956), 165–174].

75. Maximum likelihood estimation for the multinomial distribution [Sankhyā, 18, (1957), 139–148].

76. Theory of the method of estimation by minimum chi-square [Bulletin Institute International Statist., 35 (2), (1957), 25–32].

77. Some statistical methods for comparison of growth curves [Biometrics, 14, (1958), 1–17].

78. (With Dhirendra Nath Majumdar) Bengal anthropometric survey, 1945: A statistical study [Sankhyā, 19, (1958), 201–208].

79. Maximum likelihood estimation for the multinomial distribution with infinite number of cells [Sankhyā, 19, (1958), 211–218].

80. Some problems involving linear hypotheses in multivariate analysis [Biometrika, 46, (1959), 49–58].

81. Sur une caractèrisation de la distribution normale ètablie d'après une propriètè optimum des estimations linèaires (in French) [in Le Calcul des Probabilitès et Ses Applications: Paris, 15–20 Juillet 1958, Colloques Internationaux du Centre National de la Recherche Scientifique, Paris, Vol. 87, (1959), 165–171].

82. (With I. M. Chakravarti) Tables for some small sample tests of significance for Poisson distributions and 2×3 contingency tables [Sankhyā, 21, (1959), 315–326].

83. Expected values of mean squares in the analysis of incomplete block experiments and some comments based on the [Sankhyā, 21, (1959), 327–336].

84. Experimental designs with restricted randomisation [Bulletin Institute International Statist., 37 (3), (1960), 397–404].

85. Multivariate analysis: An indispensable statistical aid in applied research [Sankhyā, 22, (1960), 317–338].

86. Some observations on multivariate statistical methods in anthropological research [Bulletin Institute International Statist., 38 (4), (1961), 99–109].

87. A combinatorial assignment problem [Nature, 191, (1961), 100].

88. Asymptotic efficiency and limiting information [Proceedings of the 4th Berkeley Symposium on Mathematical Statistics and Probability: University of California, Berkeley, June 30–July 30, 1960] [(Jerzy Neyman Ed.), University of California Press, Berkeley, Vol. 1, (1961), 531–545].

89. A study of large sample test criteria through properties of efficient estimates, Part I: Tests for goodness of fit and contingency tables [Sankhyā, Ser. A, 23, (1961), 25–40].

90. A study of BIB designs and replications 11 to 15 [Sankhyā, Ser. A, 23, (1961), 117–127].

91. Combinatorial arrangements analogous to orthogonal arrays [Sankhyā, Ser. A, 23, (1961), 283–286].

92. Generation of random permutations of a given number of elements using random sampling numbers [Sankhyā, Ser. A, 23, (1961), 305–307].

93. Quantitative studies in sociology: Need for increased use in India [Sociology, Social Research and Social Problems in India, All-India Sociological Conferences: (1955–1959), (R. N. Saksena, Ed.), Asia Publishing House, Bombay, (1961), 53–74].

94. First and second order asymptotic efficiencies of estimates (with discussion) Ann. Fac. Sci. Univ. Clermont Ferrand, 8 [Actes du Colloque de Mathèmaticiens Rèuni a Clermont a l'Occasion du Tricentenaire de la Mort de Blaise Pascal, Tome II, (1962), 33–40].

95. Some observations on anthropometric surveys [Indian Anthropology: Essays in Memory of D. N. Majumdar (T. N. Madan and Gopāla Sarana), Asia Publishing House, Bombay, (1962), 135–149].

96. Efficient estimates and optimum inference procedures in large samples (with discussion) [Journal of the Royal Statistical Society, Ser B 24, (1962), 46–72].

97. A note on a generalized inverse of a matrix with applications to problems in mathematical statistics [Journal of the Royal Statistical Society, Ser B 24, (1962), 152–158].

98. Problems of selection with restrictions [Journal of the Royal Statistical Society, Ser B 24, (1962), 401–405].

99. Use of discriminant and allied functions in multivariate analysis [Sankhyā, Ser. A, 24, (1962), 149–154].

100. Apparent anomalies and irregularities in maximum likelihood estimation (with discussion) [Sankhyā, Ser. B, 24, (1962), 73–101].

101. Ronald Aylmer Fisher, F. R. S.—An obituary [Science and Culture, 29, (1962), 80–81].

102. Criteria of estimation in large samples [Sankhyā, Ser. A, 25, (1963), 180–206].

103. (With V. S. Varadarajan) Discrimination of Gaussian processes [Sankhyā, Ser. A, 25, (1963), 303–330].

104. Sir Ronald Aylmer Fisher—The architect of multivariate analysis [Biometrics, 20, (1964), 286–300].

105. (With Herman Rubin) On a characterization of the Poisson distribution [Sankhyā, Ser. A, 26, (1964), 295–298].

106. The use and interpretation of principal component analysis in applied research [Sankhyā, Ser. A, 26, (1964), 329–358].

107. The theory of least squares when the parameters are stochastic and its application to the analysis of growth curves [Biometrika, 52, (1965), 447–458].

108. Efficiency of an estimator and Fisher's lower bound to asymptotic variance (with discussion) [Bulletin Institute International Statist., 41 (1), (1965), 55–63].

109. On discrete distributions arising out of methods of ascertainment in Classical and Contagious Discrete Distributions [Proceedings of the International Symposium: McGill University, Montreal, August, 1963 (Ganapati P. Patil, Ed.), Statistical Publishing Society, Calcutta, (1965), 320–332].

110. Problems of selection involving programming techniques [Proceedings of the IBM Scientific Computing Symposium on Statistics: October 1963, IBM Data Processing Division, White Plains, New York; (1965), 29–51].

111. (With A. M. Kagan and Yu. V. Linnik) On a characterization of the normal law based on a property of the sample average [Sankhyā, Ser. A, 27, (1965), 405–406].

112. Discriminant function between composite hypotheses and related problems [Biometrika, 53, (1966), 339–345].

113. Covariance adjustment and related problems in multivariate analysis [Multivariate Analysis, Proceedings of the First International Symposium: Dayton, Ohio, 14–19 June, 1965 (Paruchuri R. Krishnaiah Ed.), Academic Press, New York, (1966), 87–103].

114. Generalized inverse for matrices and its applications in mathematical statistics [Research Papers in Statistics: Festschfift for J. Neyman (F. N. David, Ed.), Wiley and sons, London, (1966), 263–279].

115. Characterisation of the distribution of random variables in linear structural relations [Sankhyā, Ser. A, 28, (1966), 251–260].

116. (With M. N. Rao) Linked cross-sectional study for determining norms and growth curves: A pilot survey on Indian school-going boys [Sankhyā, Ser. B, 28, (1966), 231–252].

117. On vector variables with a linear structure and a characterization of the multivariate normal distribution [Bulletin Institute International Statist., 42 (2), (1967), 1207–1213].

118. Least squares theory using an estimated dispersion matrix and its application to measurement of signals [Proceedings of the Fifth Berkeley Symposium on Mathematical Statistics and Probability: Berkeley, California, 1965/1966, Vol. I: Statistics (Lucien M. Le Cam and Jerzy Neyman Eds.), University of California Press, Berkeley, (1967), 355–372].

119. On some characterisations of the normal law [Sankhyā, Ser. A, 29, (1967), 1–14].

120. Calculus of generalized inverse of matrices—I: General theory [Sankhyā, Ser. A, 29, (1967), 317–342].

121. (With C. G. Khatri) Solution to some functional equations and their applications to characterization of probability distributions [Contributions in Statistics and Agricultural Sciences: Presented to Dr. V. G. Panse on his 62nd Birthday (Govind Ram Seth and J. S. Sarma, Eds.), Indian Society of Agricultural Statistics, New Delhi, (1968), 147–160].

122. Discrimination among groups an assigning new individuals [Proceedings of the Conference on the Role and Methodology of Classification in Psychiatry and Psychopathology: Washington, D. C., 1965 (Martin M. Katz, Jonathan O. Cole and Walter E. Barton Eds.), Public Health Service Publication 1584, National Institute of Mental Health, U. S. Department of Health, Education and Welfare, Chevy Chase, Maryland, (1968), 229–240].

123. (With B. Ramachandran) Some results on characteristic functions and characterizations of the normal and generalized stable laws [Sankhyā, Ser. A, 30, (1968), 125–140].

124. (With C. G. Khatri) Some characteristics of the gamma distribution [Sankhyā, Ser. A, 30, (1968), 157–166].

125. (With C. G. Khatri) Solutions to some functional equations and their applications to characterization of probability distributions [Sankhyā, Ser. A, 30, (1968), 167–180].

126. A note on a previous lemma in the theory of least squares and some further results [Sankhyā, Ser. A, 30, (1968), 259–266].

127. (With Sujit K. Mitra) Some results in estimation and tests of linear hypotheses under the Gauss-Markoff model [Sankhyā, Ser. A, 30, (1968), 281–290].

128. (With Sujit K. Mitra) Simultaneous reduction of a pair of quadratic forms [Sankhyā, Ser. A, 30, (1968), 313–322].

129. (With Sujit K. Mitra) Conditions for optimality and validity of simple least squares theory [Annals Math. Statist., 40, (1969), 1617–1624].

130. A decomposition theorem for vector variables with a linear structure [Annals Math. Statist., 40, (1969), 1845–1849].

131. Cyclical generation of linear subspaces in finite geometries [Combinatorial Mathematics and Its Applications, Proceedings of Conference: Chapel Hill, 10–14 April, 1967 (R. C. Bose and T. A. Dowling Eds.), University of North Carolina Press, Chapel Hill, (1969), 513–535].

132. Recent advances in discriminatory analysis [Journal Indian Society of Agricultural Statistics, 21, (1969), 3–15].

133. Some characterizations of the multivariate normal distribution [Multivariate Analysis—II, Proceedings of the Second International Symposium: Dayton, Ohio, 17–22 June, 1968 (Paruchuri R. Krishnaiah Ed.), Academic Press, New York, (1969), 321–328].

134. A multidisciplinary approach for teaching statistics and probability [Sankhyā, Ser. B, 31, (1969), 321–340].

135. Inference on discriminant function coefficients [Essays in Probability and Statistics, University of North Carolina Press, Chapel Hill, (1970), 587–602].

136. Estimation of heteroscedasic variance in linear models [Journal of American Statistical Association, 65, (1970), 161–172].

137. (With B. Ramachandran) Solutions of functional equations arising in some regression problems and a characterization of the Cauchy law [Sankhyā, Ser. A, 32, (1970), 1–30].

138. Some aspects of statistical inference in problems of sampling from finite populations (with discussion and a reply by the author) [Foundations of Statistical Inference, Proceedings of the Symposium: University of Waterloo, 31st March–9th April, 1970 (V. P. Godambe and D. A. Sprott, Eds.), Holt, Rinehart and Winston of Canada, Toronto, Ontario, (1971), 177–202].

139. Estimation of variance and covariance components—MINIQUE theory [Journal of Multivariate Analysis, 1, (1971), 257–275].

140. Minimum variance quadratic unbiased estimation of variance components [Journal of Multivariate Analysis, 1, (1971), 445–456].

141. Taxonomy in anthropology [Mathematics in the Archaeological and Historical Sciences, Proceedings of Anglo-Romanian Conference: Mamaia, 1970 (F. R. Hodson, D. G. Kendall and Peter Tautu, Eds.), Edinburgh University Press. (1971), 19–29].

142. (With J. K. Ghosh) A note on some translation-parameter families of densities of which the median is an m. I. e [Sankhyā, Ser. A, 33, (1971), 91–93].

143. Characterization of probability laws by linear functions [Sankhyā, Ser. A, 33, (1971), 265–270].

144. (With Sujit Kumar Mitra) Further contributions to the theory of generalized inverse of matrices and its applications [Sankhyā, Ser. A, 33, (1971), 289–300].

145. Unified theory of linear estimation [Sankhyā, Ser. A, 34, (1972), 371–394].

146. Some comments on the logarithmic series distribution [Statistical Ecology, Proceedings of the International Symposium on Statistical Ecology: New Haven, 1969 (G. P. Patil, E. C. Pielou and W. E. Waters, Eds.), Penn State Statistics Series, Pennsylvania State University Press, University Park, Vol. 1, (1972), 131–142].

147. Recent trends of research work in multivariate analysis [Biometrics, 28, (1972), 3–22].

148. Data analysis and statistical thinking [Economic and Social Development: Essays in Honour of Dr. C. D. Deshmukh (S. L. N. Simha Ed.), Vora and Co., Bombay, (1972), 383–392].

149. Estimation of variance and covariance components in linear models [Journal of American Statistical Association, 67, (1972), 112–115].

150. (With C. G. Khatri) Functional equations and characterization of probability laws through linear functions of random variables [Journal of Multivariate Analysis, 2, (1972), 162–173].

151. (With Sujit Kumar Mitra) Generalized inverse of a matrix and its applications [Proceedings of the Sixth Berkeley Symposium on Mathematical Statistics and Probability: Berkeley, California, 1970/1971, Vol. I: Theory of Statistics, University of California Press, Berkeley, (1972), 601–620].

152. (With Sujit Kumar Mitra and P. Bhimasankaram) Determination of a matrix by its subclasses of generalized inverses [Sankhyā, Ser. A, 34, (1972), 5–8].

153. A note on the IPM method in unified theory of linear estimation [Sankhyā, Ser. A, 34, (1972), 285–288].

154. Some recent results in linear estimation [Sankhyā, Ser. A, 34, (1972), 369–378].

155. Prasanta Chandra Mahalanobis: 1893–1972 [Biographical Memoirs of the Fellows of Royal Society, 19, (1973), 485–492].

156. Unified theory of least squares [Communications Statistics, 1, (1973), 1–8].

157. (With A. M. Kagan and Yu. V. Linnik) Extension of Darmois-Skitovic theorem to functions of random variables satisfying an addition theorem [Communications Statistics, 1, (1973), 471–474].

158. (With R. Chakraborty and D. C. Rao) The generalized Wright's model [Genetic Structure of Populations, Hawaii Conference on Population Structure (Newton E. Morton Ed.), Population Genetics Monographs, 3, University Press of Hawaii, Honolulu, (1973), 55–59].

159. Representations of best linear unbiased estimators in the Gauss-Markoff model with a singular dispersion matrix [Journal of Multivariate Analysis, 3, (1973), 276–292].

160. Mahalanobis era in Statistics: A collection of articles dedicated to the memory of P. C. Mahalanobis [Sankhyā, Ser. B, 35, (1973), 12–26].

161. (With Sujit Kumar Mitra) Theory and application of constrained inverse of matrices [SIAM Journal of Applied Mathematics, 24 (1973), 473–488].

162. Some combinatorial problems of arrays and applications to design of experiments [A Survey of Combinatorial Theory, North-Holland, Amsterdam, (1973), 349–359].

163. Projectors, generalized inverses and the BLUE's [Journal of the Royal Statistical Society, Ser B, 36, (1974), 442–448].

164. (With Sujit Kumar Mitra) Projections under seminorms and generalized Moore Penrose inverses [Linear Algebra Applications, 9, (1974), 155–167].

165. Some problems of sample surveys, Suppl. Adv. in Appl. Probability 7 [Proceedings of the Conference on Directions for Mathematical Statistics: University of Alberta, Edmonton, 12–16 August, 1974 (S. G. Ghurye Ed.), (1975), 50–61].

166. Simultaneous estimation of parameters in different linear models and applications to biometric problems [Biometrics, 31, (1975), 545–554].

167. Teaching of statistics at the secondary level: An interdisciplinary approach [International Journal of Mathematics Education in Science and Technology, 6, (1975), 151–162].

168. Some problems in the characterization of the multivariate normal distribution (Inaugural Linnik Memorial Lecture) [A Model Course on Statistical Distributions in Scientific Work and the International Conference on Characterizations of Statistical Distributions with Applications: Proceedings of NATO Advanced Study Institute: University of Calgary, 29 July–10 August, 1974 (Ganapati P. Patil, Samuel Kotz and J. K. Ord, Eds.), D. Reidel, Dordrecht, Vol. 3, (1975), 1–13].

169. Growing responsibilities of government statisticians [Occasional Papers, Asian Statistical Institute (Statistical Institute for Asia and the Pacific), Tokyo, no. 4, (1975), 12].

170. On a unified theory of estimation in linear models—A review of recent results [Perspectives in Probability and Statistics—Papers in Honour of M. S. Barlett on the Occasion of his 65th Birthday, Applied Probability Trust, University of Sheffield; Academic Press, New York, (1975) 89–104].

171. Statistical analysis and prediction of growth [Proceedings of the 8th International Biometric Conference: Constanja, Romania, August 25–30, 1974 (L. C. A. Corsten and Tiberiu Postelnicu, Eds.), Editura Academiei Republicii Socialiste Romania, Bucharest, (1975), 15–21].

172. Functional equations and characterization of probability distributions [Proceedings of the International Congress of Mathematicians: Vancuver, B. C., 1974, Canada Mathematics Congress, Montreal, Quebec, Vol. 2, (1975), 163–168].

173. (With Sujit Kumar Mitra) Extensions of a duality theorem concerning g–inverses of matrices [Sankhyā, Ser. A, 37, (1975), 439–445].

174. Theory of estimation of parameters in the general Gauss-Markoff model [A Survey of Statistical Design and Linear Models, Proceedings of the International Symposium: Colorado State University, Fort Collins, 19–23 March, 1973 (Jagadish N. Srivastava, Ed.), North-Holland, Amsterdam, (1975), 475–487].

175. Characterization of prior distributions and solution to a compound decision problem [Annals Statistics, 4, (1976), 823–835].

176. Estimation of parameters in a linear model (First 1975 Wald Memorial Lecture) [Annals Statistics, 4, (1976), 1023–1037].

177. (With M. L. Puri) Augmenting Shapiro-Walk test for normality [*Contributions to Applied Statistics: Dedicated to Professor Arthur on the Occasion of his 70th Birthday* (Walter Johann Ziegler, Ed.), Experimentia Supplementum, Vol. 22, Birhaüser-Verlag, Basel, (1976), 129–139].

178. (With C. G. Khatri) Characterization of multivariate normality—I: Through independence of some statistics [Journal of Multivariate Analysis, 6, (1976), 81–94].

179. A natural example of a weighted binomial distribution [American Statist., 31, (1977), 24–26].

180. (With G. P. Patil) The weighted distributions: A survey of their applications [Applications of Statistics (P. R. Krishnaiah Ed.), North-Holland, Amsterdam, (1977), 383–405].

181. Statistics for accelerating economic and social development (Anniversary Address) [Proceedings of the 25th Anniversary of the Central Statistical Organization (1977)].

182. Cluster analysis applied to a study of race mixture in human populations [Classification and Clustering, Proceedings of the Advanced Seminar: University of Wisconsin, Madison, 3–5 May, 1976 (John Van Ryzin, Ed.), Mathematical Research Centre Publication 37, University of Wisconsin, Madison, (1977), 175–197].

183. Prediction of future observations with special reference to linear models [Multivariate Analysis—IV, Proceedings of the Fourth International Symposium: Dayton, Ohio, 16–21 June, 1975 (Paruchuri R. Krishnaiah Ed.), North-Holland, Amsterdam, (1977), 193–208].

184. Some thoughts on regression and prediction [Proceedings of the Symposium to Honour Jerzy Neyman on his 80th Birthday: Warsaw, April 3–10, 1974 (R. Bartoszynski, E. Fideli and W. Klonecki, Eds.), PWN: Polish Scientific Publishers, Warsaw, (1977), 277–292].

185. Simultaneous estimation of parameters: A compound decision problem [Statistical Decision Theory and Related Topics—II, Proceedings of 2nd Purdue Symposium: 17–19 May, 1976 (Shanti S. Gupta and David S. Moore, Eds.), Academic Press, New York, (1977), 327–350].

186. (With G. P. Patil) Weighted distributions and size-biased sampling with applications to wildlife populations and human families [Biometrics, 34, (1978), 179–189].

187. (With Nobuo Shinozaki) Precision of individual estimators in simultaneous estimation of parameters [Biometrika, 65, (1978), 23–30].

188. Least square theory for possibly singular models [Canadian Journal of Statistics, 6, (1978), 19–23].

189. A note on the unified theory of least squares [Comm. Statist., A—Theory of Methods 7, (1978), 409–411].

190. Choice of best linear estimators in the Gauss-Markoff model with a singular dispersion matrix [Comm. Statist., A—Theory of Methods 7, (1978), 1199–1208].

191. P. C. Mahalanobis [International Encyclopedia of Statistics, (William H. Kruskal and Judith M. Tanur, Eds.), The Free Press, New York, Vol. 1, (1978), 571–576].

192. Separation theorems for singular values of matrices and their applications in multivariate analysis [Journal of Multivariate Analysis, 9, (1979), 362–377].

193. Estimation of parameters in the singular Gauss-Markoff model [Comm. Statist., A—Theory of Methods 8, (1979), 1353–1358].

194. (With Haruo Yanai) General definition and decomposition of projectors and some applications to statistical problems [Journal Statist. Plann. Inference 3 (1979), 1–17].

195. (With R. C. Srivastava) Some characterizations based on a multivariate splitting model [Sankhyā, Ser. A, 41, (1979), 124–128].

196. MINQUE theory and its relation to ML and MML estimation of variance components [Sankhyā, Ser. B, 41, (1979), 138–153].
197. Matrix approximation and reduction of dimensionality in multivariate statistical analysis [Multivariate Analysis—V, Proceedings of the Fifth International Symposium: Pittsburgh, 19–24 June, 1978 (Paruchuri R. Krishnaiah Ed.), North-Holland, Amsterdam, (1980), 3–22].
198. (With R. C. Srivastava, Sheela Talwalker and Gerald, A. Edgar) Characterization of probability distributions based on a generalized Rao-Rubin condition [Sankhyā, Ser. A, 42, (1980), 161–169].
199. (With Jurgen Kleffe) Estimation of variance components [Analysis of Variance, Handbook of Statistics 1 (Paruchuri R. Krishnaiah Ed.), North-Holland, Amsterdam, (1980), 1–40].
200. A lemma on g-inverse of a matrix and computation of correlation coefficients in the singular case [Comm. Statist., A—Theory of Methods 10, (1981), 1–10].
201. (With C. G. Khatri) Some extensions of the Kantorovich inequality and statistical applications [Journal of Multivariate Analysis, 11, (1981), 498–505].
202. Obituary: Professor Jerzy Neyman (1894–1981) [Sankhyā, Ser. A, 43, (1981), 247–250].
203. Some comments on the minimum mean square error as a criterion of estimation [Statistics and Related Topics, Proceedings of the International Symposium: Ottawa, Ontario, 5–7 May, 1980 (M. Csörgö, D. A. Dawson, J. N. K. Rao and A. K. Md. E. Saleh, Eds.), North-Holland, Amsterdam, (1981), 123–143].
204. (With Jacob Burbea) On the convexity of some divergence measures based on entropy function [IEEE Transactions on Information Theory, 28, (1982), 489–495].
205. (With Jacob Burbea) On the convexity of higher order Jensen differences based on entropy functions [IEEE Transactions on Information Theory, 28, (1982), 961–963].
206. (With Jacob Burbea) Entropy differential metric, distance and divergence measures in probability spaces: A unified approach [Journal of Multivariate Analysis, 12, (1982), 575–596].
207. (With Jacob Burbea) Some inequalities between hyperbolic functions [The Mathematics Student, 50, (1982), 40–43].
208. (With Ka-Sing Lau) Integrated Cauchy functional equation and characterizations of the exponential law [Sankhyā, Ser. A, 44, (1982), 72–90].
209. (With C. G. Khatri) Some generalizations of Kantorovich inequality [Sankhyā, Ser. A, 44, (1982), 91–102].
210. Analysis of diversity: A unified approach [Statistical Decision Theory and Related Topics—III, Proceedings of 3rd Purdue Symposium: 1–5 June, 1981 (Shanti S. Gupta and James O. Berger, Eds.), Academic Press, New York, Vol. 2, (1982), 233–250].
211. Diversity and dissimilarity coefficients: A unified approach, [Theoretical Population Biology, 21, (1982), 24–43].

212. Gini-Simpson index of diversity: A characterization, generalization and applications [Utilitus Mathematic., 21 B (Special Issue dedicated to Frank Yates on the occasion of his Eightieth Birthday), (1982), 273–282].

213. (With T. Kariya and P. R. Krishnaiah) Inference on parameters of multivariate normal populations when some data is missing [Developments in Statistics, Vol. 4, Academic Press, New York, (1983), 137–184].

214. An extension of Deny's theorem and its application to characterizations of probability distributions [A Fetschrift for Erich L. Lehman, Wadsworth, Belmont, California, (1983), 348–366].

215. Optimum balance between statistical theory and applications in teaching [Proceedings of the First International Conference on Teaching Statistics: University of Sheffield, 9–13 August, 1982 (D. R. Grey, P. Holmes, V. Barnett and G. M. Constable, Eds.), University of Sheffield, Vol. 2, (1983), 34–49].

216. Statistics, statisticians and public policy making [Sankhyā, Ser. B, 45, (1983), 151–159].

217. Multivariate analysis: Some reminiscences on its origin and development [Sankhyā, Ser. B, 45, (1983), 284–299].

218. (With B. K. Sinha and K. Subramanyam) Third order efficiency of the maximum likelihood estimator in the multinomial distribution [Statistical Decisions, 1, (1983), 1–16].

219. Likelihood ratio tests for relationships between two covariance matrices [Studies in Econometrics, Time Series and Multivariate Statistics in Honour of Theodore W. Anderson (Samuel Karlin, Takesh Amemiya and Leo A. Goodman, Eds.), Academic Press, New York, (1983), 529–543].

220. Prasanta Chandra Mahalanobis (1893–1972) [Bulletin of the Mathematical Association of India, 16, (1984), 6–19].

221. (With Thomas Mathew and Bimal Kumar Sinha) Admissible linear estimation in singular linear models [Comm. Statist., A—Theory of Methods 13, (1984), 3033–3045].

222. (With Jochen Müller and Bimal Kumar Sinha) Inference on parameters in a linear model: A review of recent results [Experimental Design, Statistical Models and Genetic Statistics: Essays in Honour of Oscar Kempthorne (Klaus Hinkelmann Ed.), Statistics, Textbooks Monographs, 50, Marcel Dekker, New York, New York, (1984), 277–295].

223. (With Robert Boudreau) Diversity and cluster analyses of blood group data on human populations [Human Population Genetics: The Pittsburgh Symposium (Aravinda Chakravart, Ed.), Van Nostrand Reinhold, New York, (1984), 331–362].

224. Convexity properties of entropy functions and analysis of diversity [Inequalities in Statistics and Probability, IMS Lecture Notes Monograph Ser. 5, (1984), 68–77].

225. Use of diversity and distance measures in the analysis of qualitative data [Multivariate Statistical Methods in Physical Anthropology, (G. N. van Vark and W. H. Howell, Eds.), D. Reidel, Dordrecht, (1984), 49–67].

226. (With Jacob Burbea) Differential metrics in probability spaces [Probab. Math. Statist., 3, (1984), 241–258].
227. (With Ka-Sing Lau) Solution to the integrated Cauchy functional equation on the whole line [Sankhyā, Ser. A, 46, (1984), 311–318].
228. Inference from linear models with fixed effects [Statistics: An appraisal, Proceedings of the Conference Marking the 50th Anniversary of the Statistical Laboratory: Iowa State University, Ames, 13–15 June, 1983 (H. A. David, Eds.), Iowa State University Press, Ames, (1984), 345–369].
229. Prediction of future observations in polynomial growth curve models, [Statistics: Applications and New Directions, Proceedings of the Indian Statistical Institute Golden Jubilee International Conference on Statistics: Calcutta 16–19 December, 1981 (J. K. Ghosh and J. Roy, Eds.), Indian Statistical Institute, Calcutta, (1984), 512–529].
230. Optimization of functions of matrices with applications to statistical problems [W. G. Cochran's Impact on Statistics (Poduri S R. S. Rao and Joseph Sedransk, Eds.), Wiley and sons, New York, (1984), 191–202].

List of Publications of Professor Anadi Sankar Gupta

1. Advancement of a compressible heat-conducting fluid over an infinite flat plate [Zeit. Ang. Math. Mech., 37, (1957), 349].
2. Shear flow of a viscoelastic fluid past a flat plate with suction [Journal of Aero Space Science, (U.S.A.), 25(9), 1958].
3. Effect of buoyancy forces on certain viscous flows with suction [Applied Scientific Research (Holland), 8, (1959), 309].
4. Steady and transient free convection of an electrically conducting fluid from a vertical plate in the presence of a magnetic field [Applied Scientific Research (Holland), A, 9, (1960), 319].
5. Flow of an electrically conducting fluid past a porous flat plate in the presence of a transverse, magnetic field [Zeit. Ang. Math. Phys., 11, (1960), 43].
6. On the flow of an electrically conducting fluid near an accelerated plate in the presence of a magnetic field [Journal Physical Society Japan, 15, (1960), 1PI].
7. Laminar stagnation flow of an-electrically conducting fluid against an infinite plate in the presence of a magnetic field [Applied Scientific Research (Holland), B, 9, (1961), 45].
8. Laminar free convection flow of an electrically conducting fluid from a vertical plate with uniform surface heat flux and variable wall temperature in the presence of magnetic field [Zeit. Ang. Math. Phys., 13, (1962), 201].
9. (With L. N. Howard) On the hydrodynamic and hydromagnetic stability of swirling flows [Journal of Fluid Mechanics, 14, (1962) 463].
10. Rayleigh-Taylor instability of viscous electrically conducting fluid in the presence of a horizontal magnetic field [Journal Physical Society Japan, 18, (1963), 1073].

11. Laminar flow in plane waves of a conducting fluid in the presence of a transverse magnetic field [American Institute of Aeronautics and Astronautics, 1, (1963), 2391].

12. On the capillary instability of a jet carrying an axial current with or without a longitudinal magnetic field [Proceedings of the Royal Society (London), A, 278, (1964), 214].

13. On heat transfer characteristics of two-dimensional, circular and radial (wall) jets [Arch. Mech. Stos., Poland, 17, (1965), 547].

14. (With U. S. Rao) Hydromagnetic free convection past a vertical porous plate subjected to suction or injection [Journal Physical Society Japan, 20, (1965), 1936].

15. Tidal wave propagation in a rotating conducting fluid with a magnetic field [American Institute of Aeronautics and Astronautics, 3, (1965), 156].

16. Effect of a standing sound on the magnetohydrodynamic flow past a flat plate [Zeit. Ang. Math. Phys., 17, (1966), 260].

17. Hydromagnetic free convection flows from a horizontal plate [American Institute of Aeronautics and Astronautics, 4, (1966), 1439].

18. (With U. S. Rao) Hydromagnetic flow due to a rotating disc subjected to large suction [Journal Physical Society Japan, 21, (1966), 2390].

19. Stability of a viscoelastic liquid film flowing down an inclined plane [Journal of Fluid Mechanics, 28, (1967), 17].

20. Circulatory flow of a conducting liquid about a porous rotating cylinder in a radial magnetic field [American Institute of Aeronautics and Astronautics, 5, (1967), 380].

21. Hall effects on thermal instability [Rev. Roum. De Math. Pures et Applica., TOME XII, (5), (1968), 665].

22. Stability of a viscous liquid flowing down a flexible boundary [Canadian Journal of Physics, 46, (1968), 2059].

23. Periods of oscillation of rotating column of a perfectly conducting liquid in the presence of a uniform axial current [Progress of Mathematics, 2, (1968), 71].

24. (With Lajpat Rai) Hydromagnetic stability of a liquid film flowing down an inclined conducting plane [Journal Physical Society Japan, 24, (1968), 626].

25. Flow of a compressible radiating fluid passing infinite plate with suction [American Institute of Aeronautics and Astronautics, 6, (1968), 2209].

26. (With A. S. Chatterjee) Dispersion of soluble matter in the hydromagnetic laminar flow between two parallel plates [Proceedings of the Cambridge Philosophical Society, 164, (1968), 1209].

27. (With Lajpat Rai) Note on stability of a viscoelastic liquid film flowing down an inclined plane [Journal of Fluid Mechanics, 33, (1968), 87].

28. Combined free and forced convection effects on the magneto-hydrodynamic flow through a channel [Zeit. Ang. Math. Phys., 20, (1969), 506].

29. (With Lajpat Rai) Finite amplitude effects on magnetohydrodynamic thermal convection in a rotating layer of a conducting fluid [Journal of Mathematics, Analysis and Applications, 29, (1970), 123].

30. (With Lajpat Rai) Stability of an elastico-viscous liquid film flowing down an inclined plane [Proceedings of the Cambridge Philosophical Society, 63, (1967), 527].

31. Magnetohydrodynamic Ekman layer [Acta Mechanica, Germany, 13, (1972), 155].

32. Effect of conducting walls on the dispersion of soluble matter in MED channel flow [Rev. Roum de Physique, 15, (1970), 811].

33. (With P. S. Gupta) Asymptotic suction problem in the flow of Micropolar liquids [Acta Mechanica, Germany, 15, (1972), 142].

34. (With S. Sen Gupta) Thermohaline convection with finite amplitude in a rotating fluid [Zeit. Ang. Math. Phys., 22 Fasc., 5, (1971), 906].

35. Ekman layer on a porous plate [Physics of Fluids, U. S. A., 15, (1972), 930].

36. (With P. S. Gupta) Effect of homogeneous and heterogeneous reactions on the dispersion of a solute on the laminar flow between two plates [Proceedings of the Royal Society of London, A 330, (1972), 59].

37. Combined free and forced convection past a porous vertical circular cylinder [Indian Journal of Physics, 46, (1972), 521].

38. Decay of vortices in a viscoelastic liquid [Meccanica, Italy, VII(4), (1972), 232].

39. Thermo-convective waves in certain elastico-viscous liquids [Japanese Journal of Applied Physics, 12, (12), (1973), 1881].

40. Diffusion with chemical reaction from a point source in a moving stream [Canadian Journal of Chemical Engineering, 52, (1974), 424].

41. Hall effects on generalized MHD Couette flow with heat transfer [Bull. De L'Academie Royale de Belgique (classe de sciences), LX, (1974), 332].

42. (With B. S. Dandpath) On the stability of swirling flow in magnetogasdynamics [Quarterly of Applied Mathematics, U. S. A., 33, (1975), 182].

43. Radiation effect on hydromagnetic convection in a vertical channel [International Journal of Heat and Mass Transfer, 17, (1974), 1437].

44. (With I. Pop) Boundary layer growth in a liquid with suspended particles [Bull. Math. De la Sci. Math. Roumanie, 19, (1975), 291].

45. Hall effects on combined free and forced convective hydromagnetic flow through a channel [Letters in Applied Engineering Science, U. S. A., 14, (1976), 285].

46. Flow and heat transfer in the hydromagnetic Ekman layer on a porous plate with Hall effects [International Journal of Heat and Mass Transfer, 19, (1976), 523].

47. Hydromagnetic instability in a rotating channel flow [Publ. de L'Institute Mathematique (Belgrade), 19(33), (1975) 147].

48. On hydromagnetic flow and heat transfer in a rotating fluid past an infinite porous wall [Zeit. Ang. Math. Mech., 55, (1975), 147].

49. Hydromagnetic flow past a porous plate with Hall effects [Acta Mechanica, Germany, 22, (1975), 281].

50. Instability of a horizontal layer of a viscoelastic liquid on an oscillating plane [Journal of Fluid Mechanics, 72, (1975), 425].

51. On the non-linear stability of flow of a dusty gas [Journal of Mathematics, Analysis and Applications, 55, (1976), 284].

52. (With B. S. Majumder) Taylor diffusion in a falling film of a non-Newtonian liquid [International Journal of Heat and Mass Transfer, 20, (1977), 341].

53. (With B. S. Dandpath) Stability of magnetogasdynamics shear flow [Acta Mechanica, Germany, 28, (1977), 77].

54. (With I. Pop) Effects of curvature on unsteady free convection past a circular cylinder [Physics of Fluids, U. S. A., 20(1), (1977), 162].

55. (With P. S. Gupta) Squeezing flow between parallel plates [Wear, England, 45, (1977), 177].

56. (With N. Datta and R. N. Jana) Compressible flow past an oscillating porous plate [Japanese Journal of Applied Physics, 16(9), (1977), 1659].

57. Effect of suspended particles on the Ekman boundary layer [Analele Universitatii Bucuresti Mathematica, Roumania, XXVI, (1977), 45].

58. On the dispersion of a dye with a harmonically varying concentration in the hydromagnetic flow through a channel [Journal of Applied Physics, U. S. A., 48, (1977), 5344].

59. (With P. S. Gupta) Heat and mass transfer on a stretching sheet with suction or blowing [Canadian Journal of Chemical Engineering, 55, (1977), 744].

60. Hall effects on the hydromagnetic flow past an infinite porous flat plate [Journal of Physical Society of Japan, 43, (1977), 1767].

61. (With B. S. Dandpath) Notes on the flow near a wall and dividing streamline intersection [American Institute of Aeronautics and Astronautics, 16, (1978), 849].

62. (With B. S. Dandpath) Long waves on a layer of a viscoelastic fluid flowing down an inclined plane [Rheologica Acta, 17, (1978), 492].

63. (With A. Chakraborty) Hydromagnetic flow and heat transfer over a stretching sheet [Quarterly of Applied Mathematics, U. S. A., 37, April (1979), 73].

64. (With M. Das Gupta) Convective instability of a layer of a ferromagnetic fluid rotating about a vertical axis [International Journal of Engineering Science, U. S. A., 17, (1979), 271].

65. (With K. Rajagopal) Flow and stability of second-grade fluids between two parallel plates [Acta Mechanica, Germany, 33, (1981), 5].

66. On a boundary layer theory for non-Newtonian fluids [Letters in Applied Engineering Science, U. S. A., 18, (1980), 875].

67. (With N. Annapurna) Exact analysis of unsteady M.H.D. convective diffusion [Proceedings of the Royal Society of London, A, 367, (1979), 281].

68. (With B. S. Dandpath) Thermal instability in a porous medium with random vibrations [Acta Mechanica, Germany, 43, (1982), 37].

69. (With A. Chakraborty) Nonlinear thermohaline convection in a rotating porous medium [Mech. Res. Commu., U. S. A., 8, (1981), 9].

70. (With C. S. Yih) Plane buoyant plumes [Rev. Br. C. Mechanique, Rio de Janeiro, 111, (1981), 49].

71. (With P. Muhuri) Free convection boundary layer on a flat plate due to small fluctuations in surface temperature [Zeit. Ang. Math. Mech., 59, (1979), 117].

72. (With K. Rajagopal) On a class of exact solutions to the equations of motion of a second-grade fluid [International Journal of Engineering Science, U. S. A., 19, (1981), 1009].

73. (With K. Rajagopal) Flow and stability of a second-grade fluid between two parallel plates rotating about non-coincident axes [International Journal of Engineering Science, U. S. A., 19, (1981), 1401].

74. Free convection effects on the flow past an accelerated vertical plate in an incompressible dissipative fluid [Rev. Roum, Sci., Tech., Mech. Appl., 24, (1979), 561].

75. (With N. Annapurna) Dispersion of matter in flow of a Bingham plastic in a tube [Chemical Engineering Communications, U. S. A., 8, (1981), 281].

76. (With R. N. Jana and N. Datta) Unsteady flow in the Ekman layer of an elasto-viscous liquid [Rheologica Acta, 21, (1982), 733].

77. (With K. Rajagopal and B. S. Dandpath) On nonlinear stability of flow of a conducting fluid past a porous flat plate in a transverse magnetic field [Arch. Rational Mech and Analysis, 83, (1983), 91].

78. (With T. Y. Na and A. Nanda) Hydromagnetic flow in a channel with volume sources or sinks of mass [Journal of the National Academy of Mathematics, 1, (1983), 1].

79. (With K. Rajagopal) Remarks on 'A class of exact solutions to the equations of a second-grade fluid [Letters in Applied Engineering Science, U. S. A., 21, (1983), 61].

80. (With K. Rajagopal and T. Y. Na) A note on Falkner-Skan flows of a non-Newtonian fluid [Journal of the National Academy of Mathematics, 18, (1983), 313].

81. (With P. Muhuri) Stochastic stability of tethered buoyant platforms [Ocean Engineering, U. K., 10, (1983), 471].

82. (With K. Rajagopal and T. Y. Na) Flow of a viscoelastic fluid over a stretching sheet [Rheologica Acta, 23, (1984), 213].

83. (With K. Ganguly and S. N. Bhattacharyya) Hydromagnetic stability of helical flows permeated by a magnetic field with a radial component [International Journal of Engineering Science, U. S. A., 22, (1984), 919].

84. (With K. Ganguly and S. N. Bhattacharyya) Instability of rotating flow in magnetogas-dynamics [Journal of Mathematical and Physical Sciences, 18, (1984), 629].

85. Heat transfer in a corner flow with suction [Mech. Res. Commu., U. S. A., 11, (1984), 55].

86. (With K. Rajagopal). An exact solution for the flow of a non-Newtonian fluid past an infinite porous plate [Meccanica, Italy, 19, (1984), 158].

87. (With S. N. Bhattacharyya) On the stability of viscous flow over a stretching sheet [Quarterly of Applied Mathematics, U. S. A., XLII, (1985), 359].

88. (With S. N. Bhattacharyya) Thermoconvective waves in a binary mixture [Physics of Fluids, U. S. A., 28, (1985), 3215].

89. (With B. K. Datta and P. Roy) Temperature field in flow over a stretching sheet with uniform heat flux [International Communication of Heat and Mass Transfer, 12, (1985), 89].

90. (With N. Annapurna and B. S. Dandpath) Hydromagnetic convective diffusion between parallel plates with suction [Journal of Applied Mechanics (Trans. American Society of Mechanical Engg., U.S.A.), 52, (1985), 213].

91. (With G. Biswas and S. K. Som) Instability of a moving cylindrical liquid sheet [Journal of Fluid Engineering, U. S. A., 107, (1985), 451].

92. (With K. Ganguly) On the hydromagnetic stability of helical flows [Journal of Mathematics, Analysis and Applications, 106, (1985), 26].

93. (With K. Ganguly and S. N. Bhattacharyya) On the linear stability of hydromagnetic flow for non-axisymmetric disturbances [Journal of Mathematics, Analysis and Applications, 122, (1987), 408].

94. (With G Biswas and P. K. Nag) Heat transfer in a comer flow [Warme-und-Stoff, Germany, 21, (1987), 13].

95. (With B. K. Datta) Cooling of a stretching sheet in a viscous flow [Industrial Engg. Chemical Res., U. S. A., 26, (1987), 333].

96. (With G. Biswas) Spreading of non-Newtonian fluid drops on a horizontal plane [Mech. Res. Commu., U. S. A., 14, (1987), 361].

97. (With G. Mandal and I. Pop) Magnetohydrodynamic flow of an incompressible viscous fluid caused by axisymmetric stretching of a plane sheet [Magnitnaya Girodinamica, U. S. S. R., 23, (1987), 10].

98. (With K. Rajagopal and T. Y. Na) A non-similar boundary layer on a stretching sheet in a non Newtonian fluid with uniform free stream [Journal Mathematical and Physical Sciences, 21, (1987), 189].

99. Final stage of a falling triangular plate [Wear, England, 127, (1988), 111].

100. (With P. Ray, S. K. Bayen and B. K. Datta) Mass transfer with chemical reaction in a laminar falling film [Warme-und-Stoff, Germany, 22, (1988), 195].

101. (With S. N. Bhattacharyya) Thermoconvective waves in a rotating fluid [Physical Review, U. S. A., 38, (1988), 2440].

102. (With B. S. Dandpath) Flow and heat transfer in a viscoelastic fluid over a stretching sheet [International Journal of Nonlinear Mathematics, 24, (1989), 215].

103. (With M. K. Laha and P. S. Gupta) Heat transfer characteristics of the flow of an incompressible viscous fluid over a stretching sheet [Warme-und-Stoff, Germany, 24, (1989), 151].

104. Heat transfer in a pulsatile flow of an elastico-viscous fluid in a porous plate channel [Modelling, Simulation and Control, B, AMSE Press, 31, (1990), 1].

105. Mixed convection of an incompressible viscous fluid in a porous medium past a hot vertical plate [International Journal of Nonlinear Mathematics, 25, (1990), 723].

106. (With K, Ganguly and S. N. Bhattacharyya) On the onset of thermal instability in fluid-filled porous spheres and spherical shells [Stability and Appl. Annal. Continuous Media, 1(3), (1991), 213].

107. (With B. S. Dandpath) Stability of a thin layer of a second-grade fluid on a rotating disk [International Journal of Nonlinear Mathematics, 26, (1991), 409].

108. Hydromagnetic wave in a non-Newtonian fluid [Mech. Res. Commu., U. S. A., 19, (1992), 237].

109. Hydromagnetic stability of a stratified parallel flow varying in two directions [Astroph. Space Science, U. K., 198, (1992), 95].

110. (With H. I. Andersson, B. Holmedal and B. S. Dandpath) Magnetohydrodynamic melting flow from a horizontal rotating disk [Mathematical Models and Methods in Applied Sciences, 3, (1993), 373].

111. (With G. C. Layek, M. K. Maity and P. Niyogi) Unsteady convective diffusion in a rotating parallel plate channel [Warme-und-Stoff, Germany, 29, (1994), 425].

112. (With S. N. Bhattacharyya) Transient compressible boundary layer on a wedge impulsively set into motion [Archives of Applied Mechanics, 66, (1996), 336].

113. (With S. B. Hazra and P. Niyogi) On the dispersion of a solute in oscillating flow through a channel [Heat and Mass Transfer, Germany, 31, (1996), 249].

114. (With B. Pal and J. C. Misra) Steady hydromagnetic flow in a slowly varying channel [Proceedings of the National Academy of Sciences, 66(A), 111, (1996), 247].

115. (With B. S. Dandpath) Solitary waves on the surface of a layer of viscoelastic fluid running down an inclined plane [Rheologica Acta, 36, (1997), 135].

116. Propagation of NM thermoconvective waves [Indian Journal of Pure and Applied Mathematics, 28(5), (1997), 713].

117. (With S. B. Hazra and P. Niyogi) On the dispersion of a solute in oscillating flow of a non-Newtonian fluid in a channel [Heat and Mass Transfer, Germany, 32, (1997), 481].

118. (With J. C. Misra and B. Pal) Hydromagnetic flow of a second-grade fluid in a channel—some applications to physiological systems [Mathematical Models and Methods in Applied Sciences, 8, (1998), 1323].

119. (With S. Bhattacharyya) MH flow and heat transfer at a general three-dimensional stagnation point [International Journal of Nonlinear Mathematics, 33, (1998), 125].

120. (With S. Bhattacharyya and A. Pal) Heat transfer in the flow of a viscoelastic fluid over a stretching surface [Heat and Mass Transfer, Germany, 34, (1998), 41].

121. (With R. Deka) Flow past an accelerated horizontal plate in a rotating fluid [Acta Mechanica, Germany, 138, (1999), 13].

122. Soret effect on propagation of thermoconvective waves in a binary mixture [Heat and Mass Transfer, Germany, 35, (1999), 315].

123. (With J. C. Misra, B. Pal and A. Pal) Hydromagnetic flow of a viscoelastic fluid in a parallel plate channel with stretching walls [Indian Journal of Mathematics, 41, (1999), 231].

124. Some new similarly solutions of the unsteady Navier-Stokes equations [Mech. Res. Commu., U. S. A., 27, (2000), 485].

List of Publications of Professor (Mrs.) Jyoti Das [neè Chaudhuri]

1. On Bateman-integral functions [Mathematical Zeitschrift, 78, (1962), 25–32].
2. On a relation connecting the second solution of Tchebycheff's equations of the second order and Bessel functions [Annali Dell' Universita di Ferrara, X, (1962), 123–129].
3. A note on definite integrals involving the derivatives of hypergeometric polynomials [Rendiconti del Seminario Mathematica della Universita di Padova, XXXII, (1962), 214–220].
4. On the convergence of the eigenfunction expansion associated with a fourth-order differential equation [Quarterly Journal of Mathematics, 15, (1964), 258–274].
5. On the generalization of a formula of Rainville [Proceedings of the American Mathematical Society, 17, (1966), 552–556].
6. On the operational representation of some hypergeometric polynomials [Rendiconti del Seminario Mathematica della Universita di Padova, XXXVIII, (1967), 27–32].
7. Some special integrals [American Mathematical Monthly, 74, (1967), 545–548].
8. (With W. N. Everitt) On the spectrum of ordinary second-order differential operators [Proceedings of the Royal Society, Edinburgh, LXVIII, (1968), 95–119].
9. (With W. N. Everitt) On an eigenfunction expansion for a fourth-order singular differential equation [Quarterly Journal of Mathematics, 20, (1968), 195–213].
10. (With W. N. Everitt) On the square of a formally self-adjoint differential expression [Journal of the London Mathematical Society, (2), 1, (1968), 661–673].
11. (With W. N. Everitt) The spectrum of a fourth-order differential operator [Proceedings of the Royal Society, Edinburgh, LXVIII, (1968), 185–210].
12. (With W. N. Everitt) On the distribution of the eigenvalues and the order of the eigenfunctions the of a fourth-order singular boundary value problem [Proceedings of the Royal Society, Edinburgh, (A), 71, (1971/72), 61–65].
13. (With V. Krishna Kumar) On the eigenfunction expansion associated with a singular complex valued fourth-order differential equation [Proceedings of the Royal Society, Edinburgh, 75 A, (1975–76), 325–332].
14. (With J. Dey) On the Separation property of symmetric, ordinary second-order differential expressions [Questiones Mathematics, 1, (1976), 145–154].
15. (With J. Sett) On the Separation property of symmetric, ordinary fourth-order differential expressions [Proceedings of the Royal Society, Edinburgh, 86 A, (1980), 255–259].
16. (With M. Majumdar) On the invariance of the nature of singularities of the m-coefficients associated with an ordinary linear fourth-order differential equation [Journal of Pure Mathematics, 3, (1983), 25–30].

17. (With G. Laha) A Hilbert space associated with a singular second-order boundary value problem [Journal of Pure Mathematics, 3, (1983), 31–36].

18. (With J. Sett) A note on the separation property of symmetric fourth-order differential expressions [Journal of Indian Institute of Science, 65 (B), (1983), 173–177].

19. (With J. Sett) Non-separability of second-order differential expressions in the limit-circle case [Journal of Pure Mathematics, 4, (1983), 37–41].

20. (With M. Majumdar) On the estimation of the m-coefficients associated with a fourth-order linear differential equation [Indian Journal of Mathematics, 27, (1985), 121–129].

21. (With J. Sett) Separation and limit-classification of special fourth-order differential expression [Journal of Indian Institute of Science, 66, (1986), 547–548].

22. (With A. Chatterjee and G. Laha) Limit-classification of a real quadratic form of an ordinary second-order self-adjoint differential expression [Indian Journal of Pure and Applied Mathematics, 17 (B), (1986), 1008–1013].

23. (With P. K. Sengupta) On the point-wise convergence of the eigenfunction expansion associated with some iterated boundary value problems [Proceedings of the American Mathematical Society, 98 (4), (1986), 593–600].

24. A generalization of elliptic integrals [Festchrift for Prof. M. Datta, (1987), 73–82].

25. (With S. K. Banerjee) An alternative characterization of the spectrum of a self-adjoint operator associated with a certain fourth-order formally self-adjoint ordinary differential equation [Indian Journal of Mathematics, 30 (1), (1988), 1–8].

26. (With A. Chatterjee) On the equiconvergence of the eigenfunction expansion associated with certain second-order differential equations [Indian Journal of Pure and Applied Mathematics, 19 (11), (1988), 54–59].

27. An elementary proof of Weyl's limit-classification [Journal of the Australian Mathematical Society, Ser A, 46, (1989), 171–176].

28. (With A. K. Chakrabarty) Estimates of the eigenvalues of a mixed Strum-Liouville problem from that of the eigenvalues of a corresponding separated Strum-Liouville problem [Far East Journal of Mathematical Sciences, 2 (1), (1994), 9–16].

29. (With A. K. Chakrabarty) An alternative proof of self-adjointness condition for a mixed boundary value problem [The Mathematics Student, 64 (1–4), (1995), 9–14].

30. On the solution spaces of linear second-order homogeneous ordinary differential equations and associated boundary conditions [Journal of Mathematical Analysis and Applications, 200, (1996), 42–52].

31. (With A. K. Chakrabarty) The interlacing property of the eigenvalues of a mixed Strum-Liouville problem with the eigenvalues of a suitable separated Strum-Liouville problem [Journal of Analysis, 4, (1996), 143–151].

32. Nonlinear analysis as an aid to linear analysis [Journal of Pure Mathematics, 13, (1996), 1–12].

33. (With A. K. Chakrabarty) Summability of the eigenfunction expansion corresponding to a mixed Strum-Liouville problem [Journal of Orissa Mathematical Society, (1993–1996), 12–15].

34. (With A. K. Chakrabarty) Convergence of the eigenfunction expansion corresponding to a mixed Strum-Liouville problem [Journal of Pure Mathematics, 14, (1997), 47–55].

35. (With A. K. Chakrabarty) On the convergence of the eigenfunction expansion corresponding to a mixed Strum-Liouville problem II [Journal of Pure Mathematics, 15, (1998), 65–76].

36. (With G. Laha) Unification of Integral Transforms [Journal of Pure Mathematics, 17, (1998), 73–79].

37. A new method of solving linear homogeneous ordinary differential equations [Journal of Pure Mathematics, 17, (1998), 17–22].

38. [With M. Nandy (nee De)] The limit-point/limit-circle classification of real linear second order ordinary differential equations [Proceedings of the International Conference on Analysis and its applications held at Chennai during December 6–9, (2000), 41–47].

39. (With G. Laha) The transforms generated by a real linear regular self-adjoint boundary value problem of fourth order [Proceedings of the International Conference on Analysis and its applications held at Chennai during December 6–9, (2000), 49–61].

Printed in the United States
by Baker & Taylor Publisher Services